Tim Sandle

Produced by Microbiology Solutions.

Website: http://www.pharmamicroresources.com/

DEDICATION

This book is dedicated to my wife, Jennifer.

Data Review and Analysis for Pharmaceutical Microbiology

Dr. Tim Sandle

CONTENTS

About the author

Tim Sandle, Ph.D, M.A., BSc (Hons), CBiol, MSBiol., MIScT

Dr. Sandle is a chartered biologist (Society for Biology) and holds a first class honours degree in Applied Biology; a Masters degree in education; and obtained his doctorate from Keele University.

Dr. Sandle has over twenty years' experience as a pharmaceutical microbiologist, which includes designing, operating and reviewing a range of microbiological tests (including sterility testing, endotoxin LAL methodology, microbial enumeration, environmental monitoring, particle counting and water testing). In addition, Dr. Sandle is experienced in microbiological and quality batch review, microbiological investigation and policy development.

In addition, Dr. Sandle is an honorary consultant with the School of Pharmacy and Pharmaceutical Sciences, University of Manchester and is a tutor for the university's pharmaceutical microbiology M.Sc course. Dr. Sandle serves on several national and international committees relating to pharmaceutical microbiology and cleanroom contamination control (including the ISO cleanroom standards). He is currently chairman of the Pharmaceutical Microbiology Interest Group (Pharmig) LAL action group and serves on the National Blood Service cleaning and disinfection committee. He has written over one hundred book chapters, peer reviewed papers and technical articles relating to microbiology. Dr. Sandle has also delivered papers to over thirty conferences.

Dr. Sandle is the editor of the Pharmaceutical Microbiology Interest Group Journal and runs an on-line microbiology website and forum (www.pharmig.blogspot.com). Dr. Sandle is an experienced auditor and frequently acts as a consultant to the pharmaceutical and healthcare sectors.

Other books by Tim Sandle include:

Sandle, T. (editor) (2010): *'Current Perspectives on Environmental Monitoring: Pharmig Review Number 1'*, Published by Pharmaceutical Microbiology Interest Group, UK. ISBN 978-0-9560804-1-7

Sandle, T., Saghee, M.R. and Ramstrop, M. (2010): *Environmental Monitoring and Cleanrooms*, IDMA-APA Guideline, Technical Monograph No.5, Indian Drug Manufacturers Association, Mumbai

Saghee, M.R., Sandle, T. and Tidswell, E.C. (Eds.) (2011): *Microbiology and Sterility Assurance in Pharmaceuticals and Medical Devices*, New Delhi : Business Horizons

Sandle, T. (2011): *Two extremes? Flexible working in Europe: A study of differences in Flexible Working Time between two European Plasma Fractionators in Britain and the Netherlands*, Saarbrucken: VDM Publishing, IBSN 978-3-639-34965-8

Roberts, J. and Sandle, T. (2011). *A Guide to Microbiology Laboratories in the Pharmaceutical Industry.* Pharmaceutical Microbiology Interest Group, Pharmig: Stanstead Abbotts, UK

Sandle, T. (2012). *The CDC Handbook: A Guide to Cleaning and Disinfecting Cleanrooms*, Grosvenor House Publishing: Surrey, UK

Sandle, T. (2012). *Pharmaceutical Microbiology Glossary*, Microbiology Solutions: UK (Kindle only eBook ASIN: B0092G1246)

Sandle, T. (2012). *E-Guide to Cleanrooms*, Microbiology Solutions: UK (Kindle only eBook ASIN: B009IXFJ92)

Upton, A. and Sandle, T. (2012). *Best Practices for the Bacterial Endotoxin Test: A Guide to the LAL Assay*, Pharmaceutical Microbiology Interest Group: Stanstead Abbotts, UK

Sandle, T. and Saghee, M.R. (2013). *Cleanroom Management in Pharmaceuticals and Healthcare*, Euromed Communications: Passfield, UK

Sandle, T. (2013). *Risk Management and Risk Assessment for Pharmaceutical Manufacturing: A contamination control perspective*, Microbiology Solutions: London, UK

Sandle, T. (2013). *Sterility, Sterilisation and Sterility Assurance for Pharmaceuticals: Technology, Validation and Current Regulations*, Woodhead Publishing Ltd.: Cambridge, UK (ISBN 1 907568 38 7)

Sandle, T. (2013). *Sterility Testing of Pharmaceutical Products*, DHI /PDA: Bethesda, MD, USA (ISBN: 193372274

Chapter 1:   Introduction

The purpose of this book is to present an introduction to the types of data relating to pharmaceutical microbiology and then to present some options for the analysis and interpretation of the collected data. This involves an understanding of the data and the application of some statistical techniques. In doing so techniques and methods are presented in such a way as to allow the reader to make choices and to consider appropriate methods. For this different examples are cited and the text is illustrated by case studies.

Pharmaceutical microbiology is an important field in microbiology and relates to pharmaceutical manufacturing and healthcare. Microbiology is the — 'scientific study of the microorganisms' and is a specific branch of 'biology', which is concerned with the investigation of 'microscopic organisms' that are composed of only one cell. These are typically either unicellular or multicellular microscopic organisms that are distributed abundantly both in the living bodies of plants and animals and also in the air, water, soil, and marine kingdom.

Pharmaceutical Microbiology has a special bearing on pharmacy in all its aspects. This ranges from the manufacture and quality control of pharmaceutical products in general to an understanding of the mode of action of antibiotics. It involves the assessment of the microbiological content of drugs and medicines (which are either produced as sterile products, like liquids for injection, or as non-sterile ointments, tablets or creams) and of the environment in which medicines are made, together with an assessment of the utilities which are required for their manufacture (such as the air and water).

A typical pharmaceutical microbiology laboratory will collect thousands of sample results over the course of a year. Many of these will relate to samples collected from cleanroom environments and from water systems. The task of the laboratory is to make sense of the data and to examine the data for patterns. For this task statistics are required and techniques for data trending (there are different techniques for the examination of data for trends. This book considers several and examines two in detail: cumulative sum and Shewhart charts). In addition, the monitoring parameters – or acceptance criteria – are often calculated by the microbiologist (such as alert and action levels; with sterile operations, published alert and action levels apply to new processes. However, in time the microbiologist should develop more appropriate levels once there is sufficient historical data to work with. With non-sterile products, the microbiologist needs to develop appropriate monitoring levels from the outset).

Statistics is something which many biologists shy away from and it can be an anathema to microbiologists. This is unfortunate as statistics and data analysis can be both time saving tools and prove very useful for data interpretation. Statistics are a set of concepts, rules, and procedures that help us to:

- Organize numerical information in the form of tables, graphs, and charts;
- Understand statistical techniques underlying decisions that affect our lives and well-being; and
- Make informed decisions.

This is not undertaken simply for unlike other biological sciences pharmaceutical microbiological data does not lend itself to straightforward analysis. This is due to one of the fundamental issues of statistics, which is data distribution.

With pharmaceutical microbiology, the distribution of micro-organisms does not lend itself to statistics based on normal distribution. This is not necessarily an obstacle for data that can be transformed or, where it cannot be transformed, alternative approaches can be used.

Chapter 2: An introduction to Pharmaceutical Microbiology

Although the primary focus of this book is with microbiological data and associated statistics it is useful to clarify what pharmaceutical microbiology is and the types of samples collected. Microbiology is a biological science and is concerned with the study of microscopic organisms. Microorganisms are ubiquitous diverse, occupying almost all environments and habitats including the air, sea, land, on the human skin and inside the human body. Microorganisms include the cellular: the simple bacteria and archaea (prokaryotic) and more complex algae, fungi and protozoa (eukaryotic); and the non-cellular, viruses. Microorganisms may be both beneficial and harmful to human existence (Madigan and Markinko, 2006).

The science of microbiology is made up of several sub-disciplines, these disciplines include: Mycology (the study of fungi), phycology (the study of algae), bacteriology (the study of bacteria), parasitology (the study of parasites and virology, the study of viruses, and how they function inside cells). These broad areas encompass a number of specific fields. These fields include: immunology (the study of the immune system and how it works to protect us from harmful organisms and harmful substances produced by them); pathogenic microbiology, the study of disease-causing microorganisms and the disease process (epidemiology and aetiology); microbial genetics (which is close to molecular biology); food microbiology (studying the effects of food spoilage) and so on.

The microbiological discipline of concern to this book is pharmaceutical microbiology. Pharmaceutical microbiology is an applied branch of industrial microbiology concerned with the study of microorganisms associated with the manufacture of pharmaceuticals, primarily in controlling the numbers in a process environment; ensuring that the finished product is either sterile or free from those specific strains that are regarded as objectionable from starting materials and water, such as, *Escherichia coli*, Salmonellae, *Staphylococcus aureus* and *Pseudomonas aeruginosa*. Pharmaceutical microbiologists are also concerned with toxins (microbial by-products like endotoxins and pyrogens) ensuring that these and other 'vestiges' of microorganisms (which may elicit adverse patient responses) are absent from products.

There are different tests associated with pharmaceutical microbiology. Many of these are described in pharmacopoeias (principally the United States, European and Japanese pharmacopoeias) and Good Manufacturing Practice (GMP) guides (of which European GMP, U.S. Food and Drug Administration and World Health Organization guides are the most important). These tests include:

- Sterility Test;
- Bacterial Endotoxin Test (using *Limulus* amebocyte lysate [LAL] methodology);
- Microbial enumeration methods, primarily of starting materials ('raw materials'). This will include microbial enumeration (which is divided into tests for the total microbial count and the total yeast and mould count); presence-absence of specific indicator microorganisms (which may be considered objectionable to the product or process) and bacterial endotoxin testing;
- Antimicrobial susceptibility testing (the test measures the microbial growth response of an isolated microorganism to a particular drug or drugs);

- Methods and limits for testing pharmaceutical grade water, including Water-for-Injection (WFI) and Purified Water. Water systems are examined for Total Aerobic Microbial Count, Bacterial endotoxins and for specific microorganisms;
- Disinfectant efficacy testing;
- Pyrogen and Abnormal Toxicity Tests;
- Environmental monitoring. Environmental monitoring data indicates if cleanrooms are operating correctly, the effectiveness of cleaning and of personnel activities. Methods include air sampling (active (volumetric) air-samplers and settle plates), surface sampling (swabs and contact plates) and personnel sampling (finger dabs and gown plates);
- Bioburden testing of in-process samples or intermediate product;
- Biological indicators;
- Microbial identifications;
- Water activity;
- Microbial immersion studies;
- Cleaning validation studies;
- Maintenance of microbiological cultures;

Some of the above tests are examined in this book. The primary objective of pharmaceutical microbiology is contamination control and thus pharmaceutical microbiologists are involved in a number of aspects of the production process, utility supply and cleanroom environments. Microbiological contamination becomes a problem when it results in unwanted effects of pharmaceutical preparations caused by microorganisms or their toxic by-products. This is a concern for both sterile and non-sterile pharmaceutical products.

Although rapid methods are increasingly common in pharmaceutical microbiology the primary technique involves the use of culture media. Culture media is designed to cultivate and grow bacteria and fungi.

Chapter 3:   Microbial growth and distribution

Microbial growth requirements

Microbial growth can be defined in different ways. Growth can be defined as either an increase in both population size and population mass or simply an increase in cell number. With pharmaceutical microbiology, microbiologists tend to be most interested with increases to cell numbers (Black, 1996). With growth, bacteria grow through splitting (binary fission, as a doubling in cell number which usually occurs at the same rate that individual cells grow and divide). With cell numbers, the increase is in terms of geometric progression (which is often referred to as exponential growth). An increase in cell number is an immediate consequence of cell division. In terms of time, a bacterial generation time is also known as its doubling time (this is the time taken for a bacterium to do one binary fission starting from having just divided and ending at the point of having just completed the next division). Generation times vary with different microorganisms and different environments and can range from 20 minutes for a fast growing bacterium under ideal conditions (such as certain strains of *Escherichia coli*), to hours and days for less than ideal conditions or for slowly growing bacteria.

Bacteria grow differently in different environments. As a pure culture, bacteria grow through a series of stages described as the standard bacterial growth curve. The growth curve describes various stages of growth a pure culture of bacteria will go through, beginning with the addition of cells to sterile media and ending with a decline in viable cells. The phases of growth typically observed include:

- Lag phase (during lag phase, bacteria adapt themselves to growth conditions. It is the period where the individual bacteria are maturing and not yet able to divide. During the lag phase of the bacterial growth cycle, synthesis of RNA, enzymes and other molecules occurs).
- Exponential (or logarithmic) phase (a physiological state marked by back-to-back division cycles such that the population doubles in number every generation time. Here, the number of new bacteria appearing per unit time is proportional to the present population. Exponential growth cannot continue indefinitely, however, because the medium is soon depleted of nutrients and enriched with wastes.)
- Stationary phase (a physiological point where the rate of cell division equals the rate of cell death, hence viable cell number remains constant. The stationary phase is often due to a growth-limiting factor such as the depletion of an essential nutrient, and/or the formation of an inhibitory product such as an organic acid.)
- Death phase (exponential or logarithmic decline. At the death phase, bacteria run out of nutrients and die.)

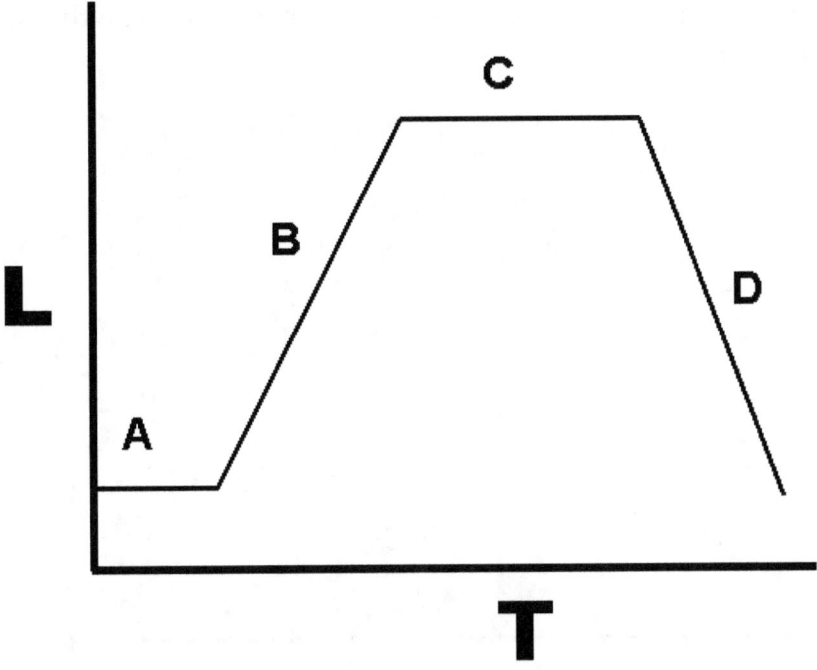

In the above picture, growth is shown as L = log (numbers) where numbers is the number of colony forming units per ml, versus T (time).

Or to illustrate growth another way, a bacterial growth curve (kinetic curve):

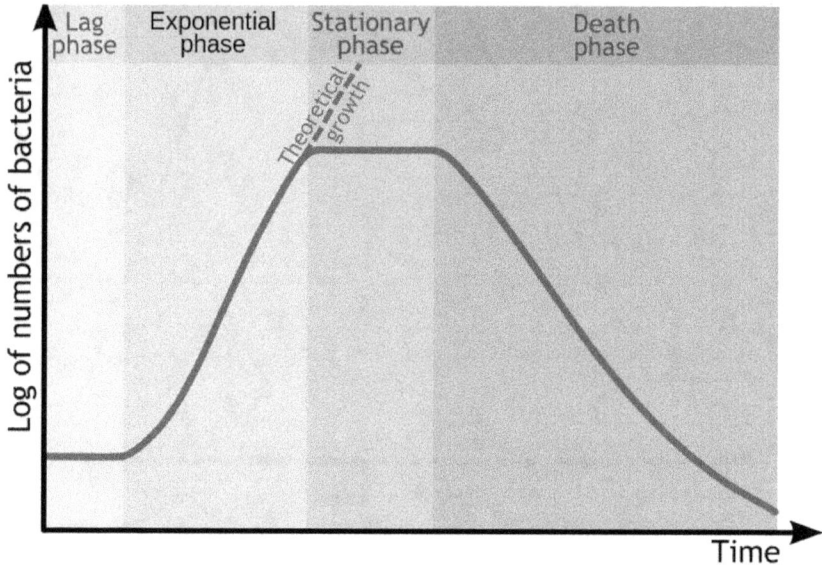

Different types of microorganisms have different growth requirements. These include:

- Growth factors and energy sources
- Temperature
- pH
- Osmolarity (Osmolarity is defined as the number of osmoles of solute per liter (L) of solution. It is expressed in terms of osmol/L or Osm/L. An osmole is a unit of measurement that describes the number of moles of a compound that contribute to the osmotic pressure of a chemical solution).
- Oxygen

These growth requirements are important considerations for the pharmaceutical microbiologist to understand as they determine the likely possibility of microbial survival and growth (Tortora *et al*, 1995).

With microbiological tests, the incubation time must be sufficiently long to account for the variable lag time characteristic of the growth curve, typical of most types of microorganism. As indicated above, growth curves begin with a lag time, the lag time is dependent upon the microorganism and how quickly the organism adapts to its new environment. Organisms which are naturally slow growing or which have been sub-lethally damaged require longer lag phases (which is why the cultural based sterility test method has an incubation time of fourteen days) (Akers, Larrimore, and Guazzo, 2003). Once the lag phase is complete, the growth phase is exponential. The growth phase is followed by death, an important consideration if the viability of the microorganism requires assessment through microbial identification.

Counting microorganisms

One of the important tasks required by pharmaceutical microbiologist is the ability to enumerate microorganisms. Counting is required in order to assess the microbial quality of water, in-process bioburden samples, of raw materials and so on. The method used to count microorganisms depends upon the type of information required, the number of microorganisms, present and the physical nature of the sample. An important distinction is between total cell count (which counts all cells, whether alive or not) and the viable count (which counts those organisms capable of reproducing).

Total cell counts include direct microscopic examination, the measurement of the turbidity of a suspension (using a nephelometer or spectrophotometer), and the determination of the weight of a dry culture (biomass assessment), ATP measurements (typically using the enzyme luciferase which produces light on the hydrolysis of ATP) fluorescent staining, or electrical impedance. Viable counting techniques include the spread plate, pour plate (through direct plating or an application like the Miles-Mistra technique), spiral plating and membrane filtration (Richter, 1999).

Sampling

The objective of taking a sample is so that the sample taken is representative of the population and by examining or testing the sample, then something meaningful can be inferred about the population. A sample is, therefore, a subset of a population selected by a process. The sample size is the number of items ('samples') included.

In terms of the numbers of samples taken (the sample design), the sample should be representative. If the sample is not representative or if the sample take is not as intended, then sampling error is said to have occurred (in practice it is very difficult to know if sampling error has occurred) (Cochran, 1977).

In terms of the sample being representative, this means that the sample should be of sufficient volume (such as 200 mL of pharmaceutical grade water) or an appropriate number of samples should be taken in order to produce a representative result (for example, determining how many samples from a give number of containers of a raw material will give a representative result. There are different statistical tools which can be used for this purpose, the most simple being the square root of the number of containers).

The reason, drawing on the water example above, that the sample size is important relates to the distribution of microorganisms. Distribution, as a general principle, is discussed in the section below. In relation to sampling, if water contains 1,000 bacteria per litre, this does not mean that each single millilitre will contain one bacterium. However, if a 500mL sample was taken, then the chance of capturing 50 bacteria is much higher than the chance of capturing 1 bacterium in a 1 mL sample. In this case, it is more useful to consider the volume required so that a reasonable estimate of the microbial population can be obtained.

For example, suppose we wish to estimate the microbial population in a litre sample. We could test the entire litre. This would itself be time consuming and expensive, and if the litre was of value, the sample would be rendered worthless.

If we want to be 95% certain of detecting a reliable count, and we know, from experience, the mean contamination level, then the following formula can be applied:

$$V = \frac{\ln (1-P)}{\mu}$$

Where:

V   =   volume of sample to test
P   =   probability of detecting an organism
μ   =   mean contamination rate

Suppose, P = 0.95 (as we are seeking 95% confidence) and the mean contamination rate is 0.022 CFU per mL. Then, the volume required would be 1,498 mL (or, in practice, a 1, 500mL sample would be tested).

In terms of where the samples are taken from, the general principle is one of random sampling. This means that the sample is selected in such a way that every sample of the same size has an equal chance of being selected. Random sampling presupposes that the population is well-mixed before sampling takes place. In drawing upon a microbiological example, random sampling would be applied to the sampling of a raw material from several drums.

Random sampling is not desirable in all cases. For sterility testing, for example, the sampling is biased. The bias is so that the samples relate to time and that the samples taken relate to approximately equal moments during the filling of the product (or, in terms of the numbers of containers filled, the samples are taken at equidistant intervals). Here, something different is being measured than if random sampling was used to select the samples.

Furthermore, an important aspect of the work of pharmaceutical microbiologists is ensuring that the samples taken or submitted to the laboratory have been done so in an aseptic manner and that the containers and storage conditions of the sample have not been adversely affected (Cundell, 2004).

Microbial distribution

It is important, before discussing general statistics and trending systems, to consider the distribution of microbial counts. This is an important topic for it has a considerable impact upon the trend charts, general statistic and on the techniques for the calculation of alert and action levels. Microbial counts in the environment rarely resemble normal distribution (where a classical bell-shaped curve or binomial pattern is obtained, where the area under the curve is divided into two symmetrical halves). Normal distribution is a phenomenon found in many aspects of physical and biological science (from measurements like human height). Normal distribution is displayed, at its most simples, as a histogram.

Before looking at microbial distribution, we need to consider distribution as a general concept for examining the patterns of data. Distributions allow data from continuous measurement scales to be summarized and statistics can be used to describe how the distribution rises and drops. The common types of data distribution are:

- Symmetric - Distributions that have the same shape on both sides of the centre are called symmetric. A symmetric distribution with only one peak is referred to as a normal distribution.
- Skewness - Refers to the degree of asymmetry in a distribution (that is, to be 'skewed' there must be a lack of symmetry). Asymmetry often reflects extreme scores in a distribution.
  Here we have:

i. Positively skewed - A distribution is positively skewed when is has a tail extending out to the right (larger numbers). When a distribution is positively skewed, the mean is greater than the median reflecting the fact that the mean is sensitive to each score in the distribution and is subject to large shifts when the sample is small and contains extreme scores.

ii. Negatively skewed - A negatively skewed distribution has an extended tail pointing to the left (smaller numbers) and reflects bunching of numbers in the upper part of the distribution with fewer scores at the lower end of the measurement scale.

- Kurtosis is a little like skewness. However, kurtosis has a specific mathematical definition, but generally it refers to how scores are concentrated in the centre of the distribution, the upper and lower tails (ends), and the shoulders (between the centre and tails) of a distribution. With kurtosis:

  o Mesokurtic - A normal distribution is called mesokurtic. The tails of a mesokurtic distribution are neither too thin nor too thick, and there are either too many or too few scores in the centre of the distribution.

  o Platykurtic - Starting with a mesokurtic distribution and moving scores from both the centre and tails into the shoulders, the distribution flattens out and is referred to as platykurtic.

  o Leptokurtic - If you move scores from shoulders of a mesokurtic distribution into the centre and tails of a distribution, the result is a peaked distribution with thick tails. This shape is referred to as leptokurtic.

In practice, the distribution of micro-organisms and microbial counts, show either Poisson distribution (such as from a water system where micro-organisms are distributed randomly) or show a marked 'skewness'. In the graph below (Figure 1), the distribution the data can be considered as skewed (that is it displays a lack of symmetry as the counts are concentrated at one end of the distribution with a long tail of counts extending in the opposite direction). This is typical of the counts obtained from an EU GMP Grade B / ISO class 7 (dynamic state) cleanroom where there are many results with zero count or counts of one or two colony forming units (cfu), and relatively few at the higher numbers. The long, thin tail towards the left of the graph can be described as negative skewness or negative exponential distribution (Wilson, 1997: 161; Everitt, 2003: 200).

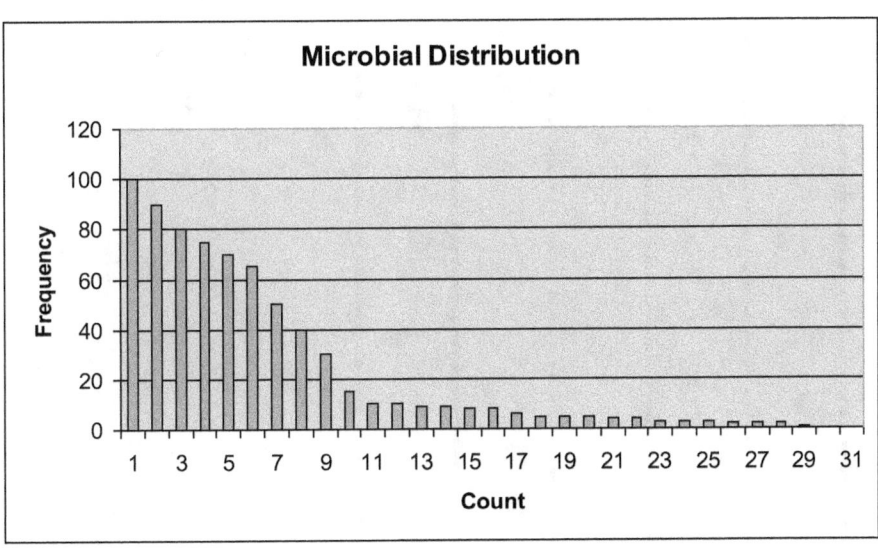

Figure 1: Illustration of negatively skewed data.

Imprecision in sampling technique can also add to this effect. This fact alone accounts for why a reasonable number of repeat samples should always be taken in response to an out-of-limits event. Such distributions can also enhance the degree of standard error obtained from plate counting.

With Poisson distribution the frequency of counting 'events' over 'time' is more random. Thus the phenomena of Poisson distribution accounts for events where a sample from a may exceed an action level on one day, be below it for another two days and then be above it again (as illustrated in Figure 2). This situation does not indicate contamination appearing and disappearing, or that one sample has given the correct result and the other an unrepresentative one.

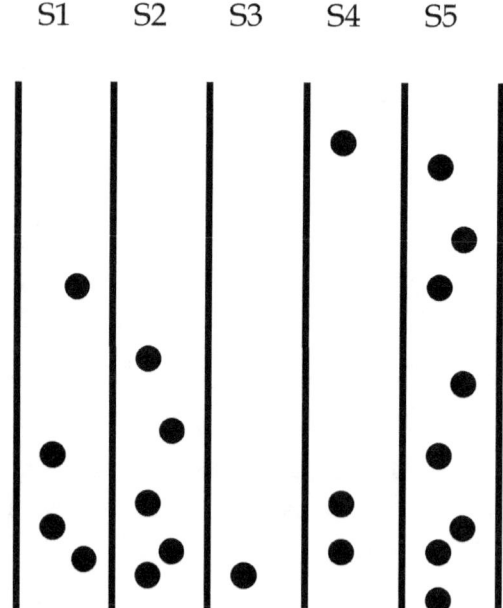

S1   S2   S3   S4   S5

Where:
Sx = sample per day
● = micro-organism

Figure 2: Depiction of typical Poisson distribution: a possible distribution of micro-organisms in five samples taken from the same location from the surface of a cleanroom on five consecutive days.

How do these observations apply to Control Charts?

The lack of normality and seeming randomness of the distribution is of importance when using control charts. Before constructing a control chart the collected data should be examined to see if it follows normal distribution. Although, as stated earlier, it is improbable that the distribution of micro-organisms as indicated by 'counts' from environmental monitoring, will follow normal distribution, any statistical analysis that is based on normal distribution remains the more accurate approach. Therefore, it is incumbent upon the user to demonstrate if there is normal distribution.

An example of such analysis is illustrated below. Table 1 summarizes the mean count from six surface locations from a Grade B cleanroom taken over a period of twenty weeks.

Table 1: Data from a Grade B cleanroom

Week No.	Mean count per week
1	0.00
2	5.15
3	0.29
4	6.93
5	1.86
6	1.47
7	0.10
8	0.00
9	2.22
10	3.95
11	0.11
12	1.25
13	0.00
14	6.34
15	0.31
16	0.45
17	2.70
18	0.89
19	0.65
20	3.45

The counts displayed are the mean cfu (colony forming units) obtained.

If the data is plotted on a blob-chart (or distribution diagram) (Figure 3), the distribution is:

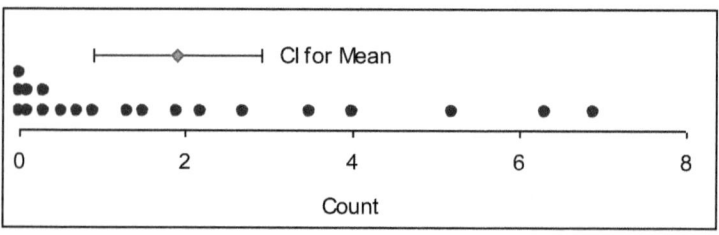

Figure 3: Distribution of typical counts from Grade B cleanroom surface over a 20 week period.

The table and distribution plot indicate that the results show a skewed distribution (as opposed to normal distribution), with the majority of the results of low count and a few higher counts to the extreme.

The data as it stands is not suitable for analysis by a control chart otherwise gross distortions would occur. It maybe that where large samples of data are taken that the so-called central limit theorem applies (that is, large samples from populations of non-normal distributions will resemble normal distribution. As such, a minimum of thirty samples will always be required) (Fleming and Nellis, 1994). This is not always the case, especially when the majority of results are low count. However, the data can be transformed into an approximation of normal distribution and thereby plotted onto a control chart. There are no pre-set 'rules' for data transformation. A common approach, as outlined by Sokal and Rohlf (1995), is to transform low count data (such as that obtained from a EU GMP Grade A environment) by taking the square root of each datum; and to transform data with higher counts by converting each datum to its logarithm at base 10 (log10). In performing log10 transformations, in the event that a count is zero, this requires the addition of a value of '1' to each value of zero. So that this does not distort relationships with other results, a value of '1' is added to each datum in the data set.

For example, to transform low count data, such as the surface count data displayed in Table 1, the square root of the mean count for the week is calculated. This is illustrated in Table 2 below.

Table 2: Surface count data (transformed)

Week Number	Mean count	Square root of mean
1	0	0.00
2	5.15	2.27
3	0.29	0.54
4	6.93	2.63
5	1.86	1.36
6	1.47	1.21
7	0.1	0.32
8	0	0.00
9	2.22	1.49
10	3.95	1.99
11	0.11	0.33
12	1.25	1.12
13	0	0.00
14	6.34	2.52
15	0.31	0.56
16	0.45	0.67
17	2.7	1.64
18	0.89	0.94
19	0.65	0.81
20	3.45	1.86

A distribution diagram of the transformed data from this table (Figure 4) shows something closer to normal distribution:

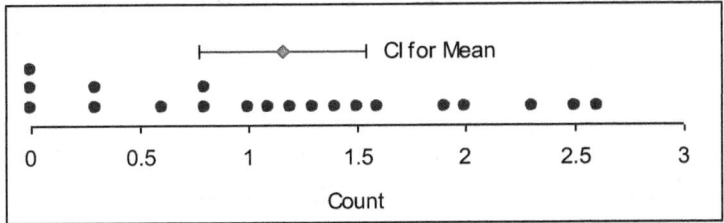

Figure 4: Revised distribution of typical Grade B surface counts over a 20 week period, based on transformed data.

For data that was of larger counts, such as from an EU GMP Grade C surface, subjecting this to logarithmic transformation would probably result in a similar change to the distribution. For example, the data in the table below has been categorised and put into a frequency table using multiples of 50. The data, as presented, does not resemble normal distribution. However, if the data is converted into logarithms and re-presented into a frequency table, the distribution of the data more closely resembles normal distribution.

When constructing frequency tables it is good practice to have ten categories or class intervals. With more than 10 intervals the table becomes cumbersome; however with too few intervals you lose information about the distribution of scores. Furthermore, the width of the interval should be a relatively simple number. For example, 2, 5, 10, or 20 would be good choices for the interval width for it is easy to count by 5s or 10s and such data is easy to understand.

An interval is usually started with the lowest score divisible by the interval size. So, if we are using a width of 10, for example, the intervals should start with 10, 20, 30, 40, etc. This makes it easier to understand the frequency table. It is important that all intervals are set at the same width. The intervals should cover the range of scores completely with no gaps and no overlaps. This means that there may be intervals within the distribution that have frequency counts of 0. This is not uncommon in distributions with extreme scores. It is also important that each score should belong to only one interval.

Displayed below are:

- Table 3: Example data set
- Table 4: Distribution of data prior to transformation
- Table 5: Data transformed to logarithms and examined for its distribution

Example data set

25	86	125	169	207	275	393
29	86	129	169	213	282	394
45	89	130	176	215	286	395
45	93	131	177	217	287	410
56	98	134	179	221	296	427
56	98	134	180	222	304	456
57	100	139	185	227	306	486
64	102	143	189	229	314	505
67	104	148	192	229	324	575
71	107	150	193	239	328	652
75	109	155	195	242	340	
76	115	162	198	247	357	
79	116	167	201	248	367	
79	117	167	205	248	370	
83	122	167	207	258	374	

Data categorised in frequency table

Mid-point	Count	Frequency
50	10	* * * * * * * * * *
100	20	* * * * * * * * * * * * * * * * * * * *
150	19	* * * * * * * * * * * * * * * * * * *
200	18	* * * * * * * * * * * * * * * * * *
250	8	* * * * * * * *
300	10	* * * * * * * * * *
350	6	* * * * * *
400	5	* * * * *
450	2	* *
500	2	* *
550	0	
600	1	*
650	1	*

Log transformed

Mid-point	Count	Frequency
3.2	1	*
3.6	2	* *
4	5	* * * * *
4.4	13	* * * * * * * * * * * * *
4.8	18	* * * * * * * * * * * * * * * * * *
5.2	25	* *
5.6	21	* *
6	13	* * * * * * * * * * * * *
6.4	3	* * *

The last table, of $\log_{10}$ transformed data, represents an alternative way of presenting data to show distribution (than the blob-chart). A further way of showing distribution would be to use a frequency histogram.

Data variations

Outside of statistical parameters it should be borne in mind that biological data is varied and there are reasons outside numerical data itself as to why microbiological data varies. Variations can arise due to several factors, including:

- The monitoring methods (which are often inherently variable)
- Culture media, where variations can arise between different types of media (such as general purpose media or fungal specific media), whether the media contains any additives, such as disinfectant neutralizer, and between manufacturers
- Incubation times
- Incubation temperatures
- Sampling procedures
- Sample size or volume
- Different sample locations
- Different sample times
- Frequency of monitoring or sampling
- The people undertaking the sampling
- Acceptance criteria (such as the means of establishing alert and action levels)

Summary

This chapter has demonstrated that the distribution of microbial counts (and of micro-organisms) rarely follows normal distribution. This presents a problem when using statistical and charting techniques that have been configured for normal distribution. The approach to remedy this is to transform the data. This is a preferable approach to using statistical techniques for Poisson distribution, which are invariably more complex.

Chapter 4: Basic Statistics

Introduction

This section outlines some of the approaches that can be taken when using statistics for interpreting microbiological data. It is one of a series of two papers which examines the application of statistics for microbiological testing. This paper is not intended to replace text-books or any other guidelines. However, there are times when different approaches and interpretations can be taken and the paper is designed as a reference source or as a point discussion. The main aim is to consider those aspects of microbiological data which do not work effectively with standard statistics. The chapter begins, however, with an overview of basic statistics.

With microbiological data, as with other scientific data, at times a hypothesis will be applied to experimental data. This is where a predicted relationship is made between two variables (a variable is some characteristic that differs from sample to sample or test to test or from subject to subject or from time to time). For example, the addition of a neutraliser to a TSA plate used for finger-dabs would be expected to result in an increase in count. A hypothesis is particularly useful for the significance test. For significance testing it is normal to use a null hypothesis and an alternative hypothesis, where:

a)   Null hypothesis is an assumption that there is no significance difference or association between the means of two sets of samples and any observed differences are simply due to chance;

b)   Alternative hypothesis is an assumption that there is a significant difference or association between the means of two sets of samples and that any observed differences are due to something other than chance.

When considering statistical techniques it should be borne in mind that statistics provides no substitute to proper and precise experimental technique or well-designed laboratory methods.

The Basics

The first part of this section considers some of the basics of statistical analysis, beginning with a discussion of data and the distribution that the data typically takes. Following this what are commonly referred to as descriptive statistics are examined. Descriptive statistics are used to describe the main features of a collection of data in quantitative terms. Descriptive statistics are distinguished from inferential statistics (or inductive statistics); in that descriptive statistics aim to quantitatively summarize a data set, rather than being used to support inferential statements.

Within the broader term 'descriptive statistics' the term 'summary statistics' is applied. Summary statistics are used to summarise a set of observations, in order to communicate the largest amount as simply as possible. Statisticians commonly try to describe the observations as:

1.   A measure of location, or central tendency, such as the arithmetic mean, median, mode, or interquartile mean.
2.   A measure of statistical dispersion like the standard deviation, variance, range, or interquartile range, or absolute deviation.
3.   A measure of the shape of the distribution like skewness   or kurtosis

Data and samples

In statistics data is considered to be representative of facts, observations, and information that come from investigations. There are broadly two categories of data:

- Measurement or observational data. This is sometimes called quantitative data, which is the result of using some instrument to measure something (e.g., test score, weight);
- Categorical data also referred to as frequency or qualitative data. Things are grouped according to some common property(ies) and the number of members of the group are recorded (e.g., males/females, vehicle type).

Microbiological data is, for statistical analysis, considered to be an observation. An observation is a measurement, either:

Qualitative	=	for example, a sterility test is either pass or a fail, or
Quantitative	=	for example, the number of colonies on a plate.

A measurement of an observation is a sample. A sample is a group of observations drawn from a population. Most microbiological data is simply the result(s) of a sample drawn from a population (such a sample of water drawn from a larger source of water). The characteristics of a population are described as parameters (for example, descriptive statistics like mean, mode and median).

Observations can be easiest seen by using a frequency table, bar graphs or histograms.

A random sample is a sample taken in a manner which gives each part of the population an equal chance of appearing in the sample. For most types of microbiological monitoring sampling is targeted (such as a location on a floor) and cannot be considered to be 'random samples'. The power of a test will be greater the larger the sample is.

Another key concept is 'variables'. A variable is the property of an object or event that can take on different values. For example, university student is a variable that takes on values like biology, computer science, English, psychology and so on.

With microbiology different classes of information are known as the variables of a dataset, e.g.:

- Colony number
- Colony size
- Colony growth rate

Variables are examined along a measurement scale. Variables can be described (classified) as nominal, ordinal, continuous, independent and dependent .

Nominal variables have a finite number of categories and the categories have no logical ordering. An example here would be the sterility test (as described in the European Pharmacopoeia chapter 2.6.1 and the United States Pharmacopoeia chapter <71>). The only possible result is 'growth' or 'no growth'.

Ordinal variables also have a finite number of categories but here the categories have a logical (often qualitative) ordering, such as the optical density of a microbial culture (eventually a concentration of microorganisms in suspension will reach its maximum density).

Continuous variables have no set number of categories and the distribution of such data tends to follow normal distribution. There are very few examples of continuous variables in microbiology and there is no clear application for pharmaceutical microbiology.

With some statistics, variables are also divided into:

- Discrete Variable - a variable with a limited number of values (e.g., gender (male/female), college class (freshman/sophomore/junior/senior).
- Continuous Variable - a variable that can take on many different values, in theory, any value between the lowest and highest points on the measurement scale.
- Independent Variable - a variable that is manipulated, measured, or selected by the researcher as an antecedent condition to an observed behaviour. In a hypothesized cause-and-effect relationship, the independent variable is the cause and the dependent variable is the outcome or effect.
- Dependent Variable - a variable that is not under the experimenter's control -- the data. It is the variable that is observed and measured in response to the independent variable.
- Qualitative Variable - a variable based on categorical data.
- Quantitative Variable - a variable based on quantitative data.

Variables which are experimentally manipulated by an investigator are called *independent variables*; whereas variables which are measured are called *dependent variables*. All other factors which may affect the dependent variable are called confounding, extraneous or secondary variables - unless these are the same for each group being tested comparisons will be unreliable.

The independent variable is typically the variable representing the value being manipulated or changed and the dependent variable is the observed result of the independent variable being manipulated. For example concerning microbial incubation, the independent variable of temperature (such as the difference between 20oC and 30oC for microbial growth) can influence the dependent variable of the number of microbial colonies recovered.

Often for many statistical investigations the data is examined for frequency distribution; that is the number of observations for each of the possible categories in a dataset. For this categories are normally listed in increasing rank order. For example, in figure 5 below:

Colony counts	Frequency:	Cumulative *f*	Cumulative %
91-100	1	50	100
81-90	9	49	98
71-80	9	40	80
61-70	13	31	62
51-60	7	18	36
41-50	6	11	22
31-40	4	5	10
21-30	0	1	2
11-20	0	1	2
1-10	1	1	2
	N = 50		

Figure 5: Distribution of colony counts using frequency distribution.

From such data, frequency histograms can be constructed. A histogram is a graphical representation of a set of observations in which class frequencies are represented as areas of rectangles placed on the class intervals (the x-axis of a graph). The height of the bars corresponds to (is proportional to) the number of observations for the particular category (as shown on the y-axis of a graph).

A pictorial representation of the data is often clearer. For example, with figure 6:

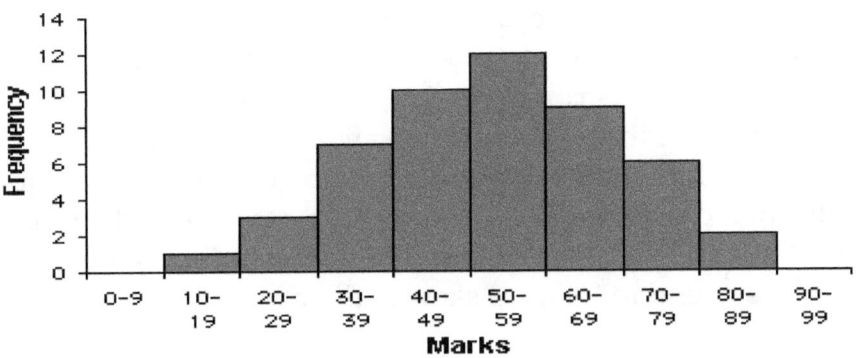

Figure 6: Illustration of a frequency histogram.

Distribution

Normal distribution is the basis of most common statistical methods. This is because:

a) Many naturally occurring populations are normally distributed;
b) The means of large random samples from populations are commonly normally distributed;
c) Many populations can be made to approximate normal distribution through data transformation (see below).

However, a problem arises because microbiological data is rarely binomial (or normally distributed). Binomial refers to the probability of an event occurring where the event has the same probability of occurring on each occasion, such as, a person being male or female. Micro-organisms in a sample follow Poisson distribution (that is the micro-organisms are not evenly distributed in any give aliquot or sub-sample of the sample, such as, a 10 mL sample from a 100 mL water sample. If the micro-organisms in a liquid are randomly distributed then they will have a Poisson distribution). Microbial counts from a test tend to follow a skewed distribution.

The testing of a sample of a liquid is significant here, for the distribution of micro-organisms on a plate will follow Poisson distribution if the bacteria were randomly distributed in the liquid that was sampled.

Normal distribution can be assessed visually using a histogram, blob chart or a normal probability plot (where the resultant plot should lie approximately along a straight line). An alternative approach is the Anderson-Darling test which determines if a sample of data has come from a population with a specific distribution (for those wishing to read more on this, refer to: Stephens, M.A. (1974): 'EDF Statistics for Goodness of Fit and Some Comparisons', *Journal of the American Statistical Association*, Vol. 69, pp730-737).

Data Transformation

Microbiological data which does not follow normal distribution may need to be transformed for certain statistical techniques and charts. A simple test of normal distribution is to use the histogram function in NWA Quality Analyst and to form a qualitative assessment. Data can also be transformed to make it easier to visualise them. Furthermore, confidence intervals and hypothesis tests will have better statistical properties if the dependent variable is approximately normal with respect to its mean, with constant variance.

Transformation is a mathematic adjustment applied to data in an attempt to make the distribution of the data fit requirements, each data point, which we will call z_i, is replaced with the transformed value:

$$y_i = f(z_i)$$

Where: f is a function.

This is undertaken because many statistical techniques are affected by the presence of skewness or outliers. In statistics this is sometimes called 'power transform'.

Considerations for use for microbiological data:

a) No one type of transformation is ideal for a particular purposes;
b) Transformations for values of 0 need to be adjusted to a minimum value of 1 (a constant will need to be added to every datum). Remember, when performing other statistical calculations, to adjust the data back by subtracting the constant;

c) For low count data (where the majority of counts are less than 10) take the square root to transform the data (alternatively squares can be taken);
d) For high count data (where the majority of counts are greater than 10) take a logarithm to the base 10. Logarithmic scales are preferred for large variations in counts. This is because; by taking the logarithm of numbers reduces the increase in count. Logarithms to the base 10 are used because:

- They are most commonly used;
- By being most commonly used, they are easiest to understand;
- For the purpose of consistency.

Taking the logarithms of numbers also makes the characteristics of data fit better and overcomes the problems associated with non-normal distribution. To assess whether normality has been achieved, a graphical approach is usually more informative than a formal statistical test. A normal quantile plot is commonly used to assess the fit of a data set to a normal population. Alternatively, rules of thumb based on the sample skewness and kurtosis have also been proposed, such as having skewness in the range of −0.8 to 0.8 and kurtosis in the range of −3.0 to 3.0.

Measures of Average (or Central Tendency)

There are three common measures of 'average': Mean, Mode and Median. These are measures of central tendency, this is a measure of the "middle", "centre" or "expected" value of the data set. There are many different descriptive statistics that can be chosen as a measurement of the central tendency of the data items. These include mean, the median and the mode. Other statistical measures such as the standard deviation and the range are called measures of spread and describe how spread out the data is.

The mean

The mean is the most common measure of central tendency. There are three types of 'mean', although only two are of concern here: arithmetic and geometric (the third, the harmonic mean, does not have a great application for most laboratory data).

Typically the arithmetic mean is used (the numerical average of the scores calculated by the sum of scores divided by the number of scores, but this may or may not be the most common score). The arithmetic mean, often simply called the 'mean', of two numbers, such as 2 and 8 cfu (colony forming units, such as from a plate count), is obtained by finding a value A such that $2 + 8 = A + A$. One may find that $A = (2 + 8)/2 = 5$ cfu. Switching the order of 2 cfu and 8 cfu to read 8 cfu and 2 cfu does not change the resulting value obtained for A. The mean 5 cfu is not less than the minimum 2 cfu nor greater than the maximum 8 cfu. If we increase the number of terms in the list for which we want an average, we get, for example, that the arithmetic mean of 2, 8, and 11 is found by solving for the value of A in the equation $2 + 8 + 11 = A + A + A$. One finds that $A = (2 + 8 + 11)/3 = 7$ cfu.

Changing the order of the three members of the list does not change the result: A (8 + 11 + 2)/3 = 7 cfu and that 7 cfu is between 2 cfu and 11 cfu. This summation method is easily generalized for lists with any number of elements. However, where numbers with decimal places are generated care must be taken. "The average colony count is 7.45 cfu" is a jarring way of making a statement in relation to the data as fragments of colonies do not exist. Here, "the average colony count is 8 cfu" would be more appropriate.

The formula for the mean is:

Mean = X = Σ Xi / n

Where:

Σ = Sum of
Xi = individual measurements
n = number of measurements

Simply, the mean is computed by summing all the scores in the distribution (SX) and dividing that sum by the total number of scores (N). The mean is the balance point in a distribution such that if you subtract each value in the distribution from the mean and sum all of these deviation scores, the result will be zero.

The mean can be very useful when comparing one sample set with another. For example, has the mean count of in-process bioburden testing changed from one month to another?

The mean is highly influenced by outliers or by data which has an asymmetric distribution. Therefore the 'mean' can be pulled away from the centre of the distribution under certain conditions:

- When the population mean is quoted this is normally designated as mu (μ).
- When the sample mean is quoted this is denoted by the lower case x bar ($\bar{x}$)

Although the sample mean is often the most commonly used for microbiological data (as an entire population can rarely be sampled it should be noted that this is arithmetic average of all the observations of the sample and not necessarily representative of the population itself). As an aside, it should be noted that both the Student's t-test and ANOVA compare means.

With the geometric mean, this is also a type of mean or average, which indicates the central tendency or typical value of a set of numbers. It is similar to the arithmetic mean, except that instead of adding the set of numbers and then dividing the sum by the count of numbers in the set, n, the numbers are multiplied and then the nth root of the resulting product is taken.

For instance, the geometric mean of the two colony counts used above, say 2 cfu and 8 cfu, is just the square root (i.e., the second root) of their product, 16, which is 4 cfu. This is slightly different to the result of 5 cfu obtained using the arithmetic mean above. The geometric mean is less affected by extreme values than the arithmetic mean and is useful for some positively skewed distributions.

The mode

Mode is the most common value occurring in a set of data (the most frequent or common score in the distribution). The most frequently occurring number in a list is called the mode. If the highest frequency is shared by more than one value, the distribution is said to be multimodal. It is not uncommon to see distributions that are bimodal reflecting peaks in scoring at two different points in the distribution.

The mode of the list of colony counts (1, 2, 2, 3, 3, 3, 4) is 3. The mode is not necessarily well defined; the list of colony counts (1, 2, 2, 3, 3, 5) has the two modes 2 and 3. The mode can be subsumed under the general method of defining averages by understanding it as taking the list and setting each member of the list equal to the most common value in the list if there is a most common value. This list is then equated to the resulting list with all values replaced by the same value. Since they are already all the same, this does not require any change. The mode is more meaningful and potentially useful if there are many numbers in the list, and the frequency of the numbers progresses smoothly. It is more usefully applied to categories rather than to numbers, such as, the typical grade of cleanroom or the most commonly occurring micro-organism from a data set.

The median

The median is the score that divides the distribution into halves; half of the scores are above the median and half are below it when the data are arranged in numerical order. The median is also referred to as the score at the 50th percentile in the distribution . Thus the median is the value which has 50% of the scores above and 50% below the value, i.e. the score in the middle of a set of data. To calculate, the data requires rank sorting in ascending or descending order. At most half the population have values less than the median and at most half have values greater than the median. The median location of N numbers can be found by the formula $(N + 1) / 2$. When N is an odd number, the formula yields a integer that represents the value in a numerically ordered distribution corresponding to the median location. (For example, in the distribution of numbers (3 1 5 4 9 9 8) the median location is $(7 + 1) / 2 = 4$. When applied to the ordered distribution (1 3 4 5 8 9), the value 5 is the median, three scores are above 5 and three are below 5. If there were only 6 values (1 3 4 5 8 9), the median location is $(6 + 1) / 2 = 3.5$. In this case the median is half-way between the 3rd and 4th scores (4 and 5) or 4.5.

It needs to be borne in mind when considering microbial colonies that there is no such thing as 'half a colony' so where a number like 4.5 is generated consideration should be considered to rounding this up to the nearest whole number, in this example '5 cfu'.

An alternative way of looking at this is if both groups contain less than half the population, then some of the population is exactly equal to the median. For example, if $a < b < c$, then the median of the list {a, b, c} is b, and if $a < b < c < d$, then the median of the list {a, b, c, d} is the mean of b and c, i.e. it is $(b + c)/2$. For further application, refer to percentiles below.

The median is not strongly influenced by outliers or asymmetry of data and it arguably, for much microbiological data, a better estimate of the typical data value than the mean.

Selecting the best measure of 'average'

With these different measures of average, where the data is symmetrical it is generally more appropriate to use the mean, whereas for skewed data or when less importance is attached to outliers, it is generally more acceptable to use the median.

The median is primarily used for skewed distributions, which it summarises differently than the arithmetic mean. Consider the multi-set of colony counts { 1, 2, 2, 2, 3, 9 }. The median is 2 in this case, as is the mode, and it might be seen as a better indication of central tendency than the arithmetic mean of 3.166.

Variability

Whereas the central tendency is a summary measure of the overall level of a dataset, variability (or dispersion) measures the amount of scatter in a dataset. For example, with figure 7:

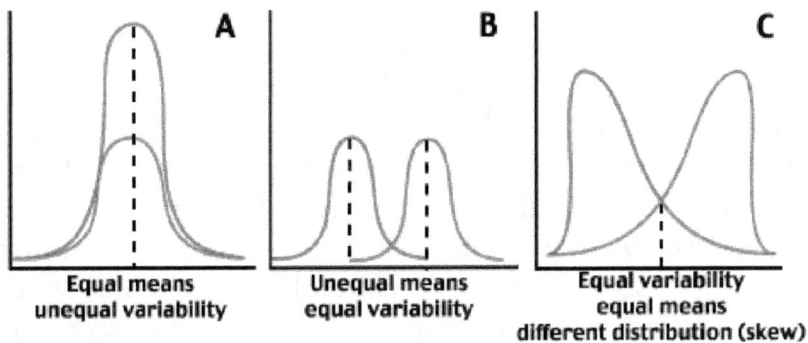

Figure 7: Diagrams illustrating data variability

Measures of spread

Although the average value in a distribution is informative about how scores are centred in the distribution, the mean, median, and mode lack context for interpreting those statistics. Measures of variability provide information about the degree to which individual scores are clustered about or deviate from the average value in a distribution.

Measures of spread are:

- Range
- Interquartile range
- Variance
- Standard deviation

Range

The simplest measure of variability to compute and understand is the range, which is the difference between the largest and smallest observation in a data set. The use of ranges (highest compared to lowest count) is useful for the examination of the variability or dispersion of microbiological data. The simplest way to express the range is to subtract the maximum value in the data set from the minimum value in the data set. Although it is easy to compute, it is not often used as the sole measure of variability due to its instability. Because it is based solely on the most extreme scores in the distribution and does not fully reflect the pattern of variation within a distribution, the range is a very limited measure of variability.

Thus the range can be easily distorted by outliers (extreme high or low scores) and it is also distorted by small sample sizes (for continuous data, as the sample size increases the minimum will probably decrease, and so the range increases). Therefore, often another criterion is used - the interquartile range (IQR). The IQR plays an important role in the graphical method known as the boxplot.

The IQR provides a measure of the spread of the middle 50% of the scores. This is the distance between the 25th and 75th percentiles (Q3 - Q1). By definition, this contains 50% of the data points in a normally-distributed dataset. The advantage of using the IQR is that it is easy to compute and extreme scores in the distribution have much less impact but its strength is also a weakness in that it suffers as a measure of variability because it discards too much data. Researchers want to study variability while eliminating scores that are likely to be accidents.

Variance

The variance measures the spread of variability (dispersion) of a sample or population from the mean. It is often expressed as the average of the squared deviations. As noted in the definition of the mean, however, simply summing the deviations will result in a value of 0. To get around this problem the variance is based on squared deviations of scores about the mean. When the deviations are squared, the rank order and relative distance of scores in the distribution is preserved while negative values are eliminated. Then to control for the number of subjects in the distribution, the sum of the squared deviations, $S(X - `X)$, is divided by n (population) or by n-1 (sample). Conventionally, n-1 is used.

The result is the average of the sum of the squared deviations and it is called the variance.

In general, the smaller the variation the greater the precision.

It is worth pausing at this point and considering the difference between the terms accuracy and precision. These, in the context of data, can be defined as:

- Accuracy – how close you are to the correct value
- Precision – how close together your results are to each other

The square root of the variance is the standard deviation (see below). The standard deviation has an advantage over the variance as by taking the square root this brings the standard deviation into the same units of measurement as that for the members of the population. Therefore, the standard deviation can be expressed as x per g or per mL etc., whereas the variance cannot.

Standard deviation

The standard deviation (s or σ) is defined as the positive square root of the variance. The variance is a measure in squared units and has little meaning with respect to the data. Thus, the standard deviation is a measure of variability expressed in the same units as the data. The standard deviation is very much like a mean or an "average" of these deviations. In a normal (symmetric and mound-shaped) distribution, about two-thirds of the scores fall between +1 and -1 standard deviations from the mean and the standard deviation is approximately 1/4 of the range in small samples (N < 30) and 1/5 to 1/6 of the range in large samples (N > 100).

In considering the application of the standard deviation for microbiological data, the standard deviation is a useful measure for the variability of microbiological data. The standard deviation is the average amount by which counts, from a set of counts, differ from the mean of that set of counts. It is most useful in providing an estimate of the variation of the counts. The standard deviation is the square root of the sample variance (with the sample variance the expression of the sum of the square differences between the sample data points and the sample data points, all divided by the sample size).

Examples of what can be inferred from the standard deviation include:

- If all of the results were identical, the standard deviation would be zero;
- The higher the standard deviation then the wider the variation.

When determining the sample size, this is expressed as n-1 to express sample variance (it should be noted that some statistical text books simply use 'n' to express population variance. This practice is not often adopted within Europe but is more commonly used in North America).

Going back to the earlier discussion on distribution, it is a common misconception that 95% of the values fall within two standard deviations of the mean. This only applies if the data is normally distributed (which, as has been shown, is not the case with very much microbiological data).

A variation of standard deviation can be used to assess whether the counts of two duplicate pour plates come from the same population. For instance:

$$\frac{\text{Highest plate count} - \text{lowest plate count}}{\sqrt{(\text{Highest plate count} + \text{lowest plate count})}}$$

Where:

Pass = <1.96
Fail = ≥ 1.96
At a 5% level of probability. The calculation is adapted from USP <1227>)

Coefficient of variation

Description: An indication of the relative variability around the mean. It can be considered as an index of precision. The standard deviation and not the variance is used to calculate the coefficient of variation because the standard deviation is expressed in the same units of measurement as the mean.

It is defined as:

100 x standard deviation / mean

In considering the application of this measure for microbiological data, some examples are:

- The coefficient of variation is frequently used in the LAL assay. Refer to rationale LR1626 for a detailed explanation of the application;
- Coefficient of variation is useful when wishing to understand the precision of a method, when the method itself is used to measure the amount of something, such as, the level of endotoxin in a sample or the microbial count;
- It can only be used for data where there is a true zero (that is data that has not been rounded to a zero).

Standard error

This is a similarly useful measure. Standard error of the mean is a measure of the extent to which the mean of a population is likely to differ from sample to sample. The bigger the standard error of the mean, the more likely the mean is to vary from one sample to another. The smaller the number of samples, the more likely it is that the sample mean will differ from the population mean. Standard error is useful when evaluating low recoveries of micro-organisms from low level challenges. For 9cm plates, the standard error increases for counts below 25 cfu (to 20% and greater). For 55mm plates (contact plates) similar standard errors are recoded with counts of 8 cfu or less.

Chapter 5: Advanced Statistics

Introduction

This section focuses on more advanced statistics. The majority of these are inferential statistics that is the use of statistics to make inferences concerning some unknown aspect of a population. There are, of course, many different statistical tests that can be considered for biological data. This section examines only some of the types of tests that can be employed.

Significance testing

In statistics, a result is called statistically significant if it is unlikely to have occurred by chance. Statistical significance is different from the standard use of the term "significance," which suggests that something is important or meaningful. To measure statistical significance a significance test is undertaken. Although it is useful for the microbiologist to know if two given data sets are or are not significantly different, significance tests are limited in that they reveal nothing about the meaning of any supposed similarity or significance of the difference.

An important consideration for any significance test is the amount of evidence required to accept that an event is unlikely to have arisen by chance. This is known as the significance level or critical p-value. For biological tests (as with many applications of significance testing elsewhere), a level of 5% (or 0.05) is chosen, for no better reason than that it is conventional. Applying the 5% level means that there is a 95 / 100 chance of being correct in the conclusion drawn. There is, however, always an element of uncertainty and importantly failing to find evidence that there is a difference does not constitute evidence that there is no difference.

All significance tests require the following approach:

- Set a hypothesis;
- Determine the alternatives to the hypothesis;
- Determine if it is a one-tailed or a two-tailed test;
- Set the significance level for the test (which is typically 5% as discussed above).

Hypothesis testing describes the procedure of assessing whether sample data are consistent or otherwise with statements made about the population.

The first step in any hypothesis testing is to state the relevant null and alternative hypotheses to be tested. This is important as misstating the hypotheses will cloud the rest of the process. The null hypothesis (H0) formally describes some aspect of the statistical "behaviour" of a set of data. This null hypothesis is assumed to be current unless the actual behaviour of the data contradicts this assumption. Here the null hypothesis is contrasted against another or alternative hypothesis.

Errors can occur with hypothesis testing. Errors are divided into Type I and Type II errors. Type I error, also known as an "error of the first kind": the error of rejecting a null hypothesis when it is actually true. It occurs when we are observing a difference when in truth there is none, thus indicating a test of poor specificity. Type II error, also known as an "error of the second kind": the error of failing to reject a null hypothesis when it is in fact not true. In other words, this is the error of failing to observe a difference when in truth there is one, thus indicating a test of poor sensitivity.

In medicine, the term Type III error is sometimes applied. An example would be identifying the poorer of two treatments as the better (Everitt, 2006).

Test: t-test

A t-test is a statistical hypothesis test in which the test statistic follows a Student's t distribution if the null hypothesis is true. The t- test is a test to determine whether two means are significantly different from each other. The t-test is one of the most commonly used significance test in microbiology. Generally, the larger the t-value the more likely the test is to be statistically significant. Considerations for the use of the t-test are:

Difference between one-tailed and two-tailed t-tests

The one-tailed is used where there are grounds for predicting the direction of the difference (such as media x will give a higher recovery than media y). Where there are no such grounds the two tailed test is applied. The t-value generally has to be higher for a two-tailed test than with a one tailed test. In most circumstances the two-tailed test is used because it is unknown whether one test are will give a higher recovery than another.

Signs (+ or -)

In most circumstances, for microbiological results, the signs are arbitrary because they depend on the use to which it is being put. For the t-test signs are ignored.

Paired or related samples

The data sets analysed by the t-test can be paired or unpaired sample sets. The unpaired, or "independent samples" t-test is used when two separate independent and identically distributed samples are obtained, one from each of the two populations being compared. Dependent samples (or "paired") t-tests typically consist of a sample of matched pairs of similar units, or one group of units that has been tested twice (a "repeated measures" t-test).

For paired samples, the t-test is to determine whether the means of two samples that come from the same or similar cases are significantly different from each other. Generally, the larger the t-value then the more likely the two means differ significantly from each other. An example of this would be water from the same outlet tested on two different media.

For unpaired or unrelated samples, the t-test is used to determine whether the means of two samples that consist of different or unrelated cases differ significantly from each other. Again, generally, the larger the t-value then the more likely the two means differ significantly from each other. An example of this would be a comparison of two air-samplers. Here the air-sampler cannot test exactly the same quantity of air as the other.

It is preferable, where possible, to use the paired t-test (that is testing the same sample to compare different methods) otherwise the chance of introducing sampling error increases.

One sample t-tests

One sample t-test can be used to compare a set of data against the sample specification. This test is generally more limited in value than t-test which compares two sets of data.

Level of significance

The level of significance is normally 5%, unless justified (see significance level below).

Interpreting t-tests

To interpret a t-test a t table is required. T-tests can be calculated manually or using a software programme. It is recommended that a software programme, such as NWA Quality Analyst, is used for accuracy and for consistency.

When using t-test tables, it is important to use the appropriate level of significance (normally 5% or p=0.05) and the appropriate column for a two-tailed or one-tailed test (some t-tables only list the two tailed test, in such circumstances the table value should be divided in two to produce the appropriate value for the one-tailed test). A p=0.05 is standard for biological data (Everitt and Palmer, 2011).

The table value should correspond to the degrees of freedom (that is the number of observations or samples included in the test. The term 'degrees of freedom' is commonly used in statistics, it essentially means the number of independent units of information in a sample) . This is normally derived from n-1 (although care should be taken as there are differences between methodologies used in the USA and in Europe, with the USA sometimes using 'n'. Where the software package NWA Quality Analyst is used, this is n-1).

The table value thus produces a numerical value. In summary, the questions to be asked are:

a)   What is the required level of significance?
b)   Is the one-tailed or two-tailed column appropriate?

c)   What are the degrees of freedom?

The table value is then compared with the calculated value of t. This produces one of two outcomes:

i.   If the calculated value of t is less than the table value it is not significant and thus there is no significant difference between the two data sets.
ii.   If the calculated value of t is more than or equal to the table value it is significant and thus there is a significant difference between the two sets of data compared.

In each case the confidence with the above outcome relates to the level of significance chosen.

At the 5% level, if the two data sets are seen to be significantly different then there is a 5% certainty that the differences are due to chance and a 95% certainty that the differences are not due to chance.

Data distribution

Like many statistical tests, t-tests have an in-built assumption that the data is normally distributed. With microbiological data, this is not always the case. In such circumstances consideration should be given to transforming the data (such as taking the square root or log to base 10).

Test: Analysis of Variance (ANOVA)

The t-test is used to analyse one factor between two groups. Analysis of variance (ANOVA) is used to examine the means of three or more groups and to determine the extent to which they differ. The ANOVA provides a statistical test of whether the means of several groups are all equal, and therefore generalizes Student's two-sample t-test to more than two groups.

What ANOVA does is separate the variation attributable to one factor from that attributed to others.

Whilst ANOVA can demonstrate whether test means differ, it cannot, however, determine in which way the means differ. This requires the comparison of two means at a time (t-test) or a combination test. Examples include comparing between counts of different bacteria; different disinfectants; or between different technicians.

Where more than one type of dependent variable are analysed at the same time multivariate analysis of variance or multivariate analysis of covariance are used (Multivariate analysis is a generic term for the many methods of analysis used to examine multivariate data; this is data for which each observation consists of values recorded onto several variables. For example, with a microorganism this might be colony shape, colony size, and colony pigmentation). There are several different models of ANOVA. General considerations for the use of ANOVA include:

a) One-way ANOVA: for only one factor or independent variable that is being measured;

b) Two-way ANOVA is where two or more factors (sometimes called factorial analysis of variance) are being examined (Factorial designs a ways to look at two or more questions in an investigation).

c) The use of ANOVA can be complicated. It is preferable to use the t-test where possible.

Some different types of ANOVA include:

- One-way ANOVA is used to test for differences among two or more independent groups. Typically, however, the one-way ANOVA is used to test for differences among at least three groups, since the two-group case can be covered by a t-test.
- One-way ANOVA is often used for repeated measures; for example, using the same micro-organisms for each treatment (such as challenging the same micro-organism against different disinfectants).
- Factorial ANOVA is used when the experimenter wants to study the effects of two or more treatment variables. The most commonly used type of factorial ANOVA is the 22 ("two by two") design, where there are two independent variables and each variable has two levels or distinct values. Factorial ANOVA can also be multi-level such as 33, etc. or higher order such as 2×2×2, etc. but analyses with higher numbers of factors are rarely done by hand because the calculations are lengthy.
- Mixed-design ANOVA. When an experimenter wishes to test two or more independent groups subjecting the subjects to repeated measures, one may perform a factorial mixed-design ANOVA, in which one factor is a between-subjects variable and the other is within-subjects variable. This is a type of mixed-effect model.
- Multivariate analysis of variance (MANOVA) is used when there is more than one dependent variable.

Probability

Probability is the mathematical chance or likelihood that a particular outcome will occur. Probabilities are used as part of significance testing but have no other useful application for microbiological data analysis. Probabilities are expressed as falling between 0 and 1. An impossible event has a probability of 0, and a certain event has a probability of 1.

Probability can be defined in different ways. One common way is in terms of frequency. For example, to assess the probability of Z:

Probability (Z)   =   number of times Z occurs
                          Number of times Z could occur

Goodness of Fit

The preferred goodness of fit test is often the Chi-squared test (χ^2 test). This is sometimes called Pearson's Chi-squared and is a measure of the independence between two variables (rather it t tests a null hypothesis that the frequency distribution of certain events observed in a sample is consistent with a particular theoretical distribution). The test examines the squared differences between the observed and expected values or frequencies.

With the Chi-squared test a comparison is made between observed frequencies to the frequencies expected by chance. An example of its application is where the test can be used for an annual assessment of pipetting accuracy for individual technicians. Chi-squared is used because it measures if a pattern of obtained frequencies differs significantly from an expected pattern of frequencies. It is a measure of variance. A 5% probability is used due the relative imprecision with the technique and occasional 'clumping' by the test bacterium.

The approximation to the chi-square distribution breaks down if expected frequencies are too low, so care must be taken to use an appropriately large sample size.

Confidence interval

A confidence interval (CI) for a population parameter is an interval between two numbers (intervals are used to indicate the reliability of an estimate). It provides an estimate of where the true (or 'population') mean falls based on an examination of the sample mean. A confidence interval is always qualified by a particular confidence level, usually expressed as a percentage; thus one speaks of a "95% confidence interval". A 95% confidence interval implies that if the process was repeated multiple times, then 95% of the calculated values would be expected to contain the 'true' value.

The end points of the confidence interval are referred to as confidence limits. For a given estimation procedure in a given situation, the higher the confidence level then the wider the confidence interval will be.

For example, expressing something as a 95% confidence interval means that there is a 95% certainty that the population mean falls between the two values quoted. This is useful when applying to validation studies as it can be stated that if the experiment was repeated several times there would be a 95% certainty of the subsequent values falling within the 95% confidence interval.

The 95% confidence limits are usually calculated from:

Estimate ± 1.96 (standard error of the estimate)

Where 1.96 is equivalent to the 97.5th percentile (for normal / Gaussian distribution), or within two standard deviations of the mean. For real life data this could mean:

- 95 of every 100 normal patient's results would be within +/- 2 S of the mean

- 1 of every 20 controls could be out of range and that is to be expected – the analytical run would be rejected

In relation to control charts and the Westgard Rules (discussed later) this rule is called the 12s rule and gives a high level of false rejections or false alarms.

Considerations for use with microbiological data include:

a) The 95% confidence interval tends to be used for microbiological data. Normally, the 95% confidence interval is considered as a greater 'certainty', such as a 99% confidence interval, cannot be reasonably applied to microbiological data;
b) Confidence intervals can be applied to sampling;
c) Confidence intervals can be assessed when comparing between duplicate results so that result A can be compared with result B to determine if the interval covering the mean for 95% of the likely samples drawn from the population. An application of this would be comparing two membrane filtration results for a sample tested in duplicate;
d) Typically a large amount of data is required to increase the 'confidence' that the quoted confidence interval is accurate.

Linear Correlation

Correlation (often measured as a correlation coefficient, ρ; correlation infers a 'co-relation') indicates the strength and direction of a linear relationship between two random variables. The correlation coefficient is an index which quantifies the linear relationship between two variables. Thus an index of the linear (straight line) relationship between two variables which can be observed. The range is -1.00 through 0 to 1.00. Note: this is different to other approaches to correlation, such as polynomial.

In interpreting correlation:

- A value of zero indicates that there is no linear relationship between two variables.
- A positive correlation is when high scores on one variable go with the high scores of the other variable;
- A negative correlation is when the high scores on one variable go with the low scores of the other variable;
- The bigger the size of the correlation, regardless of it being negative or positive, the stronger the linear relationship between the two variables (or, the closer the correlation is to 0 the lower the probability of a relationship, with a value of 0 indicating no relationship). This can be shown using a scatter diagram, with a regression line (a straight line drawn through the points on the scatter diagram) to show the linear relationship between two variables.

A scatter diagram is a two-dimensional plot of a sample with bivariate observations. The diagram indicates what kind of relationship links the two variables plotted.

- Pearson's correlation coefficient sometimes described as Pearson's product moment correlation coefficient) (r) is often used for normally distributed data. It is calculated by multiplying the standardized scores of two variables. It is vulnerable to extreme scores or outliers. Therefore, for microbiological data the use of Spearman's rank order correlation coefficient (p) is preferable. This requires the data to be ranked.
- Due to the relative imprecision of microbiological data, the following relationships are generally acceptable:

0.80 or more	=	a large, strong or high relationship
0.30 or less	=	small, weak or low relationship.
0.30 to 0.80	=	moderate or modest relationships.

a) Squaring Pearson's correlation coefficient – r2 – provides the coefficient of determination. This gives a clearer indication of the meaning of the size of a correlation as it gives the proportion of the variance between two variables.
b) The correlation can also be converted into a t-score to provide the statistical significance of the correlation.

Correlations can be expressed on scatter diagrams. These are often more meaningful than data tables. When constructing such charts the independent variable should always be on the horizontal line of the chart. For example, if the count from a sample was being examined over time, then time would be the independent variable because the objective would be to see what effect time has on the count.

With correlations the conventional dictum is that "correlation does not imply causation", which means that correlation cannot be used to infer a causal relationship between the variables. This should always be borne in mind.

Outliers

An outlier is an 'extreme' and unrepresentative value which, if retained in the data set, would have an undue and misleading influence on the interpretation of the data (that is an observation that is numerically distant from the rest of the data). Outliers can occur by chance in any distribution, but they are often indicative either of measurement error or that the population has a heavy-tailed distribution. In the former case one wishes to discard them or use statistics that are robust to outliers, while in the latter case they indicate that the distribution has high kurtosis and that one should be very cautious in using statistical techniques that assume a normal distribution. A higher-than-usual value should alert the analyst to a possible outlier.

Outliers can have many anomalous causes. A particle counter for taking measurements may have suffered a transient malfunction or there may have been an error in data transmission or transcription. Outliers arise due to changes in system behaviour, human error, instrument error or simply through natural deviations in populations. A sample may have been contaminated with elements from outside the population being examined. Alternatively, an outlier could be the result of a flaw in the assumed theory, calling for further investigation by the researcher.

Outliers have a greater impact on some statistical tests than others, for example, the calculation of the correlation coefficient is heavily affected by outliers. The use of scatter diagram provides the best check for outliers when confirming the acceptance of a correlation coefficient.

The question of rejection of a value within a data set is a delicate one. It is certainly inappropriate to replace a value simpler because we do not like it. However, it can be equally as wrong to include a value that is clearly erroneous. For example, if particle counts are collected and the particle counter has not been correctly purged, then the first value collected may well be an outlier (indicative of a count generated within the counter, rather than something reflective of the air volume sampled by the counter). A second example is the use of the wrong dilution or solution when conducting an assay (Hewitt, 2004).

There is no accepted outlier test for microbiological data although there are different statistical tests like Grubbs' test (also known as the maximum normed residual test). Such tests are based on the data distribution being normal, which poses problems for microbiological data. Nonetheless, professional judgment may be just as effective as any test, provided it is supported by a justification. The impact of outliers can also be reduced by transforming the data.

It might be that removing an outlier requires the replacement of the removed value by another. This will relate to the desired sample size and what the data was intended for (and also to how easier or difficult it is to replace the missing value, which is something which cannot be done from 'real-time' sampling such as monitoring the environment at a specific point in time). With something like an agar diffusion assay, where the Latin Square design is used, the use of a substitute value is essential to facilitate the calculation, even if the value itself does not add any information to the obtained data.

For certain biological tests, the pharmacopeias provide accepted formulae for the establishment of replacement values. For example, the following formula can be used for the replacement of a missing value (termed y') from a Petri-dish assay:

$$y' = \frac{fT_r' + kT_t' - T'}{(f-1)(k-1)}$$

Where:

F	=	number of sets (assay plates)
K	=	number of treatments
T_r'	=	the incomplete total from the plate having the missing value
T_t'	=	the incomplete total from the treatment having the missing value
T'	=	the incomplete totals from the assay as a whole

Percentiles

A percentile is the value of a variable below which a certain per cent of observations fall. When a set of scores is ranked in order from lowest to highest, the 10th percentile is the score which cuts off the bottom 10% of the scores (the smallest scores). The 50th percentile is also known as the median. Percentiles are a way of expressing a result in terms of its relative value to other results. For example, a count of 25 cfu is relatively meaningless. Whereas, knowing a count of 25 cfu is equal to the 95th percentile of a data set indicates that it is a very high result from the data set.

A related term is the percentile rank. The percentile rank of a score is the percentage of scores in its frequency distribution which are lower than it. For example, a test result which is greater than 75% of the results set is said to be at the 75th percentile. There is no 100th Percentile because no result can be lower than itself.

Quartiles are percentiles at 25th, 50th and 75th values and are used to show the measure of spread, normally given by the distance between the first and third quartiles. The difference between the 1st quartile or lower quartile (25th percentile) and the 3rd quartile or upper quartile (75th percentile) is called the interquartile range. Quartiles are normally expressed as Q2, Q3 etc. The median corresponds to the second quartile. The big advantage of using the interquartile range is that is less affected by extreme values than if the complete range was assessed.

Percentiles are used most frequently as one of the methods for the calculation of warning and action levels (see Lovegrove-Saville and Perry, 2000). Outside of this application, percentiles are relative limited because they do not give an indication of the distribution of the data or the relative frequency of occurrence.

Productivity ratio

The productivity ratio is designed to show the relative size of two numbers. It is expressed as:

Productivity ratio =   Mean of two test plate
                         Mean of two comparative control
                         plates

Using this equation a count of 12 and a count of 8 cfu would be expressed as 1.5 rather than 12:8 or 12/8.

The productivity ratio is used most frequently to assess he recovery between two different types of culture media challenged with the same micro-organism. The acceptance criteria are often taken from the Ph. Eur. (currently this is 50-200%, or when expressed as the productivity ratio: between 0.5 and 2.0).

Sample numbers

It is noted that small sample sizes will affect validity of the data. Normally most statistical results are meaningful when 100 or more samples are examined. This should normally be sought; however it is acknowledged that some process and products are limited e.g. minor products so in some circumstances 10 results may be used with appropriate justification.

A note on microbiological data

With any biological test it is good practice to examine the raw data. Such an assessment can provide important information, at first sight, as to whether the data is of the quality expected, and therefore, suitable for processing and further analysis. This is a 'common sense' inspection. When doing so, some questions worth asking are:

- Is the variation of responses within a reasonable range?
- Is the data within the analyst's general experience of the method?
- Are there any apparent outliers?
- Does the data immediately suggest a practical error, such as mis-dilution or the testing of the wrong solution?

A note on rapid methods

With the advent of rapid microbiological methods, methodological considerations associated with more long-standing biological assays become applicable to microbiological tests (Jimenez, 2004). Such considerations include:

- Sensitivity
- Accuracy
- Robustness
- Specificity
- Linearity
- Detection limit
- Robustness
- Ruggedness
- Range
- Precision

The applicability of each of the above validation parameters will depend upon the particular method and its limitations.

Summary

This section has examined further applications of statistics, drawing upon examples of microbiological data. The purpose was to offer some insights as to how microbiological data might be better interpreted and to point out some of the assumptions that need to be made and pitfalls which can occur when using statistics for biological data.

Statistics has a role within microbiology. Since microbiology involves experimentation and reaching conclusions based on inherently uncertain data, then statistics provides methodology for design and analyses of these biological tests or experiments, and hopefully more meaningful conclusions can be drawn.

Having argued the importance of statistics, it is also important to note that, when attempting to describe data, there is often no better way than to use the "interocular test," that is, to display the data visually in a graphical form. A lot of information can be gathered simply by looking at the data.

Chapter 6:   Use and calculation of warning and action levels

The use of alert and action levels and the setting monitoring limits

This section examines the regulatory requirement to set and to assess warning (or alert) levels (and sometimes action levels) on a periodic basis based on an examination of historical data. With environmental monitoring, warning (or alert) and action levels are not specifications. They are indicators of change and are used for trending purposes and for initiating investigations as required.

Based on PDA Technical Report Number 13 (2001), a workable definition of alert and action levels are:

- "Alert level: a level, when exceeded, indicates that the process may have drifted from its normal operating condition. This does not necessarily warrant corrective action but should be noted by the user".

In addition to this definition, it can be added that a warning level excursion does not adversely affect product quality but it is an indication of a potential adverse trend and a mechanism for early warning.

- "Action level: a level, when exceeded, indicates that the process has drifted from its normal operating range. This requires a documented investigation and corrective action."

Again, in addition to the PDA definition it can be added that an action level is something that could potentially affect product quality and therefore requires some action to be taken in conjunction

Note: This recommendation is also made by the UK based PHSS technical Monograph Number 2 'Environmental Control and Practice'.

Thus alert and action levels are essential tools for the examining microbiological data.

There are two approaches for alert and action levels. The first is the long established classical approach of having fixed values where sample results below the value are considered to be satisfactory and sample results at or above the value are considered to be excursions. For example, the action level for Grade B / ISO class 7 active air-samples is 10 cfu / m3. Ignoring trend monitoring for the time being, if our results are below the value of 10 they are satisfactory. If our results are at or above 10 they are considered to be unsatisfactory and an 'action' is expected (normally an investigation). This approach is that inferred by the EU GMP Guide and the FDA Guide to Aseptic Filling.

An alternative approach, which was put forward in the Pharmacopoeial Forum, Vol. 31, Issue 2, (the journal for the development of United States Pharmacopoeial chapters) in relation to the USP cleanroom chapter <1116>, is to use a frequency cut-off approach for aseptic filling operations. The rationale for this approach is that setting alert levels of <1 or zero CFU, with action levels at 1 or 2 CFU is scientifically incorrect because neither air sampling technologies nor 'conventional' or 'rapid alternatives (this includes biophysical methods such as spectrophotometric particulate analysis) microbiological methods support these requirements. For instance there are no standard methods for air sample collection and variability is as previously mentioned comparatively high and there is no data on limit of detection of environmental sampling methods (zero does not mean absence of contamination it merely means below the level of detection at that point in time).

Furthermore, it is not scientifically correct to establish recommended test criteria without a careful consideration of metrology and analytical capability and here both the limits of detection and quantization of growth and recovery methods are unknown. This is complicated yet further by sampling itself being reliant upon aseptic technique, which makes the origin of low level contamination nearly impossible to know with certainty: was a count of 1 CFU due to chance or does it reflect one microorganism?

Another complication is that at very low recovery levels there is no agreed way to establish alert or action Levels statistically because the counts are simply too low to make statistical analysis useful. On this basis it a count of one CFU is not significantly different from a count of ten CFU (indeed some scientific literature suggests that +/-0.5 log is a reasonable assumption of variability).

The alternative approach is for aseptic processing areas to be assessed on the basis of the frequency of distribution of counts and for an 'action' to be set if the level of incidences (or 'contamination events') exceeds a certain level. There is an expectation that contamination rate events for aseptic processing should be infrequent.

In a presentation by Akers (then Chairman, USP Microbiology & Sterility Assurance, Committee of Experts) to the USP, the following 'contamination rate recovery targets' were proposed:

Grade	Active air sample	Settle Plate (9cm) 4hr exposure	Contact Plate or Swab	Glove or Garment
Isolator or Closed RABS (ISO 5 or better)	<0.1%	<0.1%	<0.1%	<0.1%
ISO 5	<1%	<1%	<1%	<1%
ISO 6	<3%	<3%	<3%	<3%
ISO 7	<5%	<5%	<5%	<5%
ISO 8	<10%	<10%	<10%	<10%

Table 6: Table showing summary of the experimental data, together with the average (mean) percentage reductions.

This section considers the classical approach, as this is the current regulatory expectation, and the alternative approach (which can yield useful data and may become established as a conventional approach).

Why calculate warning and action levels?

For pharmaceutical microbiology, unlike some other disciplines such as chemistry, there are very few published specifications. Where specifications exist in microbiology these tend to be for tests where the expected outcome is straightforward, such as, pass or fail or presence-absence. These tests include the sterility test and tests for specific pathogens (both of which involve making a qualitative assessments: pass or fail).

For other important applications of pharmaceutical microbiology, such as environmental monitoring, there exist only vague 'recommended levels' or 'guidance values' that should not be exceeded (for example, the EU GMP Guide, 2009, and the FDA Guide to Aseptic Filling, 2004). It is important to note at this point that for environmental monitoring, water testing and the like we are using the term 'level' rather than 'limit'. Limit is a more absolute term implying pass or fail, accept or reject. This is appropriate for the sterility test, but meaningless for environmental monitoring where the trend is of prime importance rather than an individual count (which is rarely of significance, due to the variability in microbial counting and the effect of standard error. This is increasingly important when the action level values are small). Alert and action levels are therefore used to detect shifts from the norm and to indicate if an individual result or process is potentially out-of-control ('norm' here refers to what is usual; care needs to be taken that it is not simply used for what is desirable). This then prompts the user to:

- Assess any risk
- To propose any corrective action
- To propose any preventative action

In assessing a risk, reference should be made to the highest published level ('the specification') or to any product licence level; to the direction of the trend; and to the type of micro-organism (which can be of greater significance than the obtained count).

It is uncertain how the 'recommended levels' published in the EU GMP Guide and FDA Guide (2004) have been derived at and there are few published scientific rationales as to why, for instance, an active air-sample count at EU GMP Grade B / ISO class 7 with a count of 9 cfu / m^3 is not a risk and whereas one of 11 cfu / m^3 is a risk (based on a recommended, average maximum value of 10 cfu/m^3). One of the few attempts at this is by Tang, 1998, which examined the microbial limit test.

This curiosity aside, the published regulatory levels are only intended to apply to new processes (such as a newly built clean room). Following a period after the new process has been established, it is expected that these monitoring levels are re-assessed based on a review of operational performance. In setting localised alert and action levels, the regulatory levels should not be exceeded without a very strong rationale and being prepared to argue the case with an inspector. Or, more simply: do not ever attempt to set limits above those endorsed by regulatory agencies.

Therefore, aside from the published maximum values, regulators expect each organisation to calculate warning (or 'alert') levels based on both historical data and using some type of statistical technique (refer to the following: USP <1116>; FDA Guide to Aseptic Filling, 2004; The Gold Sheet, January 2005; PDA Technical report Number 13, 2001 Ackers, J. (1997); Deschenes, P.; Lovegrove-Saville, P. and Perry, M. (2000); Hetroys, P. et al (1997): 'Moving towards an microbiological environmental monitoring programme'; Wilson, J. (2001); EN BS ISO 14698 Parts 1 and 2; EST-RP-CC023.1; IEST-RP-CC027.1; Wilson, J. (2001); Clontz, L. (2004); PIC/S (2004) This is indeed this author's experience of FDA and MHRA inspections, and such an approach meets the requirements of the EU GMP which requires "appropriate alert and action limits" to be set.

With action levels, there is some debate as to whether these should be set based on historical data or whether the regulatory limits should be set. With environmental monitoring data, for example, both EU GMP and the FDA Guide have a recommended level of 10 cfu /m^3 for active air-samplers in Grade B / ISO class 7 (in-use) environments. It is uncertain whether it is only the alert level which requires calculating in this case or whether both an alert level and an action level are calculated, with the 10 cfu /m^3 maintained as a specification (or a third tier of action level).

This section proceeds on the basis that both alert and action levels are calculated based on historical data. This approach is useful for where, as with non-sterile environmental monitoring, there are no published recommended limits which have widespread acceptance, and the approach can be adopted for aseptic manufacturing environments.

How are alert and action levels to be calculated?

In considering the classical approach (fixed, numerical limits) there are no pre-set rules for the calculation of warning and action levels. The precise techniques and quantities of data to be used will depend upon several factors, which may include:

- The length of time that the facility has been in use for;
- How often the user intends to use the limits for (i.e. when the user intends to re-assess or re-calculate the limits. Is this yearly? Two yearly? And so on).;
- Custom and practice in the user's organisation (e.g. is there a preferred statistical technique?)

The principles behind the calculation of warning and action levels are:

- They be calculated from an historical analysis of data ;
- This should use some type of statistical technique.

These two factors are examined in turn.

Historical data

The quantity of historical data to be used is something the user will need to define. This can be time based or for a set minimum number of samples. The data selected must be representative and a minimum of one year is normally used as this will account for any seasonal variations. Seasonal variation is a term which describes cycles within a time series, which, by definition is one year or more.

To look at data meaningfully, a reasonably larger number of observations is also required in order for the data set to be representative. So, it is recommended that any analysis is, as a minimum, one year or 100 results.

If a 'special cause' event has taken place the user may decide that the affected portion of the data is unrepresentative and chose to exclude it. This should be done using pre-defined criteria, such as an outlier test, or with appropriate justification. Although there are some statistical techniques that are available, professional judgement should also be brought into use. The most straightforward approach is to plot the data set that is intended to be used to perform the calculations and note any variations. For example, if counts were typically higher during one month then possibly a seasonal factor is in play, or if counts were higher when a new technician was performing an analysis then the data may not be representative of all laboratory staff. Unrepresentative data could lead to inappropriately high action levels being calculated.

It is important to consider this because the data set must be representative. Otherwise there is a danger of applying precise statistical methods to what is already relatively imprecise and variable data (as based on the different microbiological techniques used to collect the data).

Statistical techniques

Statistical techniques are commonly divided into parametric and non-parametric techniques. The difference here is that parametric refers to a procedure which sets out to test a hypothesis about a parameter within a population described by a certain distributional form, which is typically normal distribution. Therefore, parametric methods only really apply to data sets which are normally distributed. An example of a parametric technique is Student's t-test.

Three statistical techniques for assessing monitoring levels are considered in this section: percentile cut-off; the normal distribution and negative exponential distribution approaches. The criteria for selecting these are that they are based on the 2001 PDA guide to environmental monitoring (which is, arguably, the industry's leading guidance document on this subject), and that they have some familiarity to the author. These should not be considered the only techniques available. Alternative approaches include the non-parametric tolerance limit approaches (or at a tolerance limit of $\gamma = 0.95$ and $P = 0.95$) and Weibull distribution.

For the illustration of each method, data was collected over a period of time (such as one year) and levels were set. The approach assumes that the previous period was 'normal' and that future excursions above the limits are deviations from the norm. Therefore, selecting appropriate data is in many ways more important than the choice of analytical technique.

With these considerations noted, the three techniques are examined in turn:

Percentile cut-off approach

For low count data (such as Grade A / ISO class 5 or B environmental monitoring) a percentile cut-off approach is most suitable (and the easiest to understand and to use). Percentiles are sets of divisions that produce exactly 100 equal parts in a series of continuous values. In order to calculate percentiles the data must be collected, sorted and ranked from lowest to highest.

In selecting percentile cut-off values, typically the warning level is set at the 90th or 95th percentile and the action level set at the 95th or 99th percentile. Thus if the 90th percentile is selected, this means that any result above the 90th percentile is 90% higher than values typically collected over the past year (or whatever the data selection period was). There is no easy guide as to the appropriate percentile to select. In this author's experience it is more common to use the 95th percentile for the warning level and the 99th percentile for the action level. This selection is based on the level of risk that the user wishes to build into the system. A further consideration is whether the microbiologist wishes to round up or down to the nearest zero or five. By doing so, this may make it easier for those using the levels to implement them. Laboratory staff might find it easier to recall and recognise alert and action levels of 10 and 20, as opposed to 8 and 22.

Where data is of a broad range an alternative approach is to group the data into categories using frequency distribution. For example, 0-10; 11-20; 21-30 and so on. When the category closest to the percentile cut-off is selected either the mid-point of the category or the upper value of the category can be selected. For example, if the 21-30 category represents the 95th percentile, the action level selected maybe 25 or 30 depending upon the pre-defined criteria adopted.

As an alternative, where the data exhibits extreme skewness it may be prudent to use Poisson distribution tables (Wilson, 1997: 161).

Worked example

The following data set was recorded from an environmental monitoring session. The data relates to counts obtained from active air-sampling. The object is to calculate the 95th and 99th percentile as our warning and action levels. For the illustration, 100 results are included in the sample set.

The sample set is:

72	5	10	25	45	32	61	12	18	44
52	12	56	43	40	45	67	89	23	33
44	51	55	10	6	72	34	33	21	28
29	33	49	67	71	80	33	20	16	18
44	52	89	91	32	45	41	48	37	49
33	21	19	37	86	44	47	32	39	27
21	45	42	30	27	29	27	29	12	18
33	46	76	12	30	45	64	62	34	25
27	33	38	17	32	28	34	42	51	61
59	68	72	35	27	20	14	33	47	51

Figure 8: Data set for worked example

To calculate the 95th and 99th percentile best way is to copy the data into MS Excel and to use the 'PERCENTILE' function.

This is:

=PERCENTILE(array,p)

Where the array is the data range (column G) and p is the cumulative probability (00.95 and 0.99). In this example, the 95th percentile is 76.2 (more realistically rounded down to 76 CFU) and the 99th percentile is 89.02 (or 89 CFU). This is a relatively quick calculation to undertake.

Standard deviations / Negative exponential distribution

For higher count data (such as active air-sample counts at Grade D) either standard deviations (if there is normal distribution) or negative exponential distribution (for skewed data) are employed. Unlike the percentile cut-off approach, this technique uses the mean count and observes the spread (or variance) of the different observations. For these approaches the warning level is equivalent to two standard deviations and the action level to three standard deviations of the mean.

Where the data appears normally distributed, or if a successful data transformation step such as taking the square root or logarithm has been employed, standard deviations can be used to calculate monitoring levels (although there is a danger that inaccuracies can creep in). A common mistake is to produce a histogram and use the second (for the alert level) and third (for the action level) standard deviations. This is incorrect because in doing so this approach has an inherent two-tail probability built into it. Our concern is with data moving in one direction (we don't have results below zero). Therefore if standard deviation is to be used then 2.326 + the mean CFU (for equivalence to the 99th percentile) and 1.645 + the mean CFU (for equivalence to the 95th percentile) should be used (PDA, 2001: 8-10).

The concern with this approach is that most microbiological data is not normally distributed because microorganisms in the environment (and in water) are not randomly distributed. With higher count data there is a greater chance of 'normal' distribution.

For data transformation, consider a set of environmental monitoring data for which a histogram has been constructed:

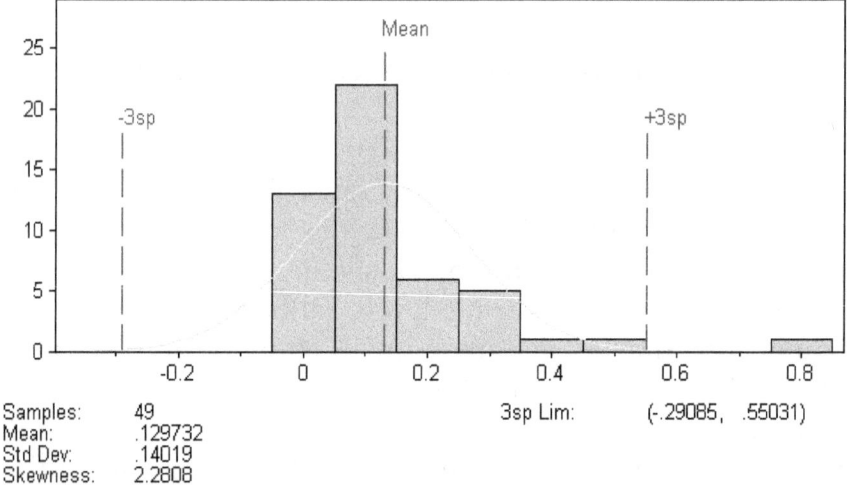

Figure 9: Histogram showing environmental monitoring data

The data is clearly not normally distributed. In attempting to transform the data by taking the square root (as the data in this example is of a relatively low count), the histogram changes to:

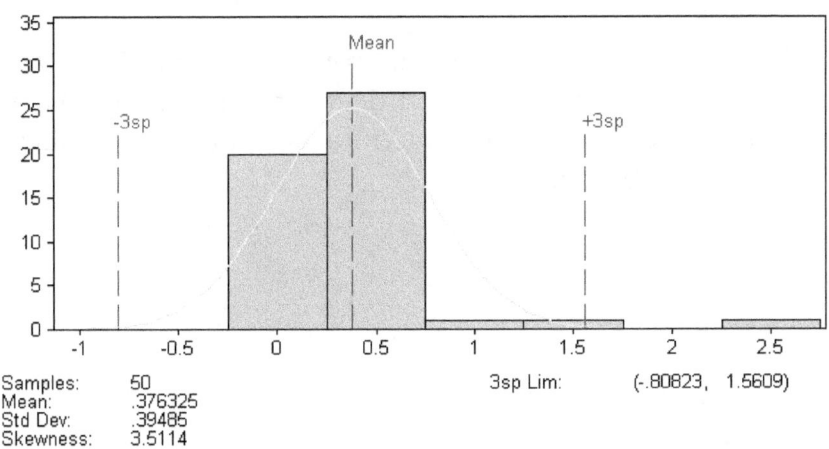

Figure 10: Histogram showing the same data, only this time transformed.

Despite an attempt to transform the data is still does not conform to normality due to the extreme values (which would distort our calculations). In this example, the percentile cut-off approach would be the best approach to take.

Negative exponential distribution (a term for negatively skewed data) provides reasonable approximation of normal distribution. In everyday life, the negative exponential distribution is often used to model real world events which are relatively rare, such as the occurrence of earthquakes; the time until a radioactive particle decays, or the time between clicks of a Geiger counter; the time it takes before the next telephone call; the time until default (on payment to company debt holders) in reduced form credit risk modelling.

By multiplying the mean by 4.6 an approximation of the 99th percentile is produced and by multiplying the mean by 3.0 an approximation of the 95th percentile is produced (Wilson, 1997: 161; PDA, 2001:10).

Post-calculation test

A useful post-calculation test (or 'sore thumb' activity) is to examine the data set as a histogram (refer to figure 5 below). If the method selected is considerably wrong, then examining where the calculated levels fall on a histogram will be obvious and distribution used to calculate the monitoring levels does not match the distribution of the actual data.

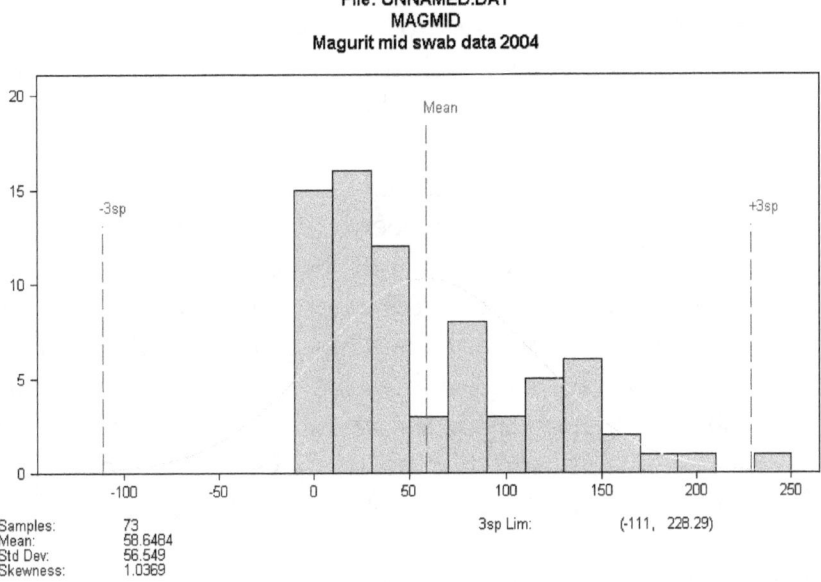

Figure 11: Histogram to illustrate distribution of microbiological data.

Problems

There are a number of problem areas associated with the calculation of warning and action levels. These include:

- Selection of an insufficiently small set of data so that the norm of the process was not captured.
- The set of data was large but for different reasons special causes resulted in it not being typical over a longer period of time.
- Whether levels should be lowered or increased following each review if the data set indicates a change in direction. Here it needs to be firmly established if any change in the historical data is due to a special cause or a common cause. If the conclusion is special cause, and these have been corrected, the monitoring levels should probably not be changed. Whereas, if changes are due to a common cause the monitoring levels should probably be changed. Even if a common cause is established it may be prudent not to change unless the trend appears over two years. The purpose of reviews should not be to drive limits upwards or downwards without very strong reasons.

It must not be forgotten that by setting monitoring levels based on the premise that 95 and 99 per cent of the data falls within, and that 5 and 1 per cent of the data falls without. Therefore, occasional excursions from these levels are to be expected for the data that is gathered and trended over the next year and some action level excursions will always be expected if the data set used for the calculations was truly representative.

Use of alert and action levels

It should be noted that for several practitioners the relevance of having an alert limit is questioned, especially when applied to non-sterile manufacturing. In many instances, if an excursion occurs it will usually go from a 'normal level' to action without an increased trend to the alert level.

Setting of monitoring limits

In contrast to sterile manufacturing, here are no major regulatory standards for the setting of limits for non-sterile manufacturing.

The setting of monitoring limits for non-sterile manufacture may involve an assessment of:

- The chemical composition of the product
- The production process
- The route of application
- Intended use of the product
- The delivery system of the product
- The type of anti-microbial preservative

When setting limits, many of the principles for sterile manufacturing are relevant.

Chapter 7: Trending microbiological environmental monitoring data

Introduction

The reason for requiring the use of trend charts relates to one of the biggest challenges in pharmaceutical microbiology: that is the vast quantity of data that is collected and the difficulty in interpreting it; and because a single monitoring event only provides a 'snap shot' of what may have been happening on a particular day, which may or may not be representative.

Thus, trend analysis is very important for environmental monitoring and is necessary in order for the microbiologist to see 'the big picture'. There is no right or wrong approach for selecting a trend analysis system, although there are dangers in selecting a trend system that is inappropriate for the data set. In this section two, systems that are familiar to the author are explored in detail, and a third approach is referenced. The two approaches explored in detail are both control charts and use quantitative assessments. Control charts are useful tools for visualizing a process and provide quick "summaries" of process statistics. Control charts make an assumption that the data plotted is independent, that is a given value is not influenced by its past value and will not affect future values. Control charts have been used for many years as part of Statistical Process Control (SPC) systems.

Reporting

Whichever method of reporting is selecting, environmental monitoring data must be presented, interpreted and summarised so that senior management can understand the trend and the 'big picture'. This data needs to be presented at the correct frequencies (that is not too often or too infrequently). There is no right or wrong way to present data. However, a clear and simple approach is often the most useful. This can include:

- The use of graphs and tables. This allows the trend of one area to be compared with another and for informed questions to be asked.
- Each test should be reported separately. Multi-line graphs and the use of more than one scale on a graph is generally confusing.
- Focus on each filling room or main operation separately. It is often useful to compare different areas but it is confusing to attempt this on one graph.
- Include all of the available data. It is important to select the time period over which the data should be collected and plotted (typically this will be monthly, quarterly, yearly or, occasionally, over a longer term). Once a time period has been selected data must never be excluded.
- Include warning and action limits on graphs. A trend can sometimes be misleading. It is important to understand how a trend relaters to the monitoring levels applied.
- Included appropriate information with tables and graphs. This helps to identify patterns and possible reasons for a given trend. Such information includes:

i.   Locations;
ii.   Dates;
iii.   Times;
iv.   Identification results;
v.   Changes to room design;
vi.   Operation of new equipment;
vii.   Shift or personnel changes;
viii. Seasons;
ix.   HVAC problems (e.g. an increase in temperature).

Displaying data

The three approaches detailed in this section are: histograms, cumulative sum charts and the Shewhart charts. These are not the only approaches that can be taken and the methods applied in this section are by no means definitive. The reason for exploring three different approaches is to emphasise their similarities (with the control charts) and differences; and, also, to demonstrate that different types of data will require different types of trending. This section explores this by using viable environmental monitoring data from two different clean rooms . One clean room is EU GMP Grade B / ISO class 7 (ISO Class 7, dynamic state) and the other is an EU GMP Grade C (ISO Class 9, in the dynamic state). Due to the difference cleanliness status of the two clean rooms the data gathered differs in the magnitude of the counts.

Histograms

Before embarking on a review of more complex control charts, there are alternative approaches that can be taken which are simpler and can capture 'the picture' as well as more sophisticated forms of analysis. The most useful of these is the frequency chart, which is particularly applicable to low count data (such as that found at Grade A / ISO class 5). Typical data, for settle plates from a filling machine examined over one year, could resemble:

Frequency category (colony forming units recovered)	Number of observations (settle plates recording counts)
0	1,120
1	35
2	7
3	2
4	0
5	1
6	0
7	0
8	0
9	0

Table 7: Frequency of counts from a Grade A / ISO class 5 cleanroom obtained over one year.

Such data can be placed into a frequency histogram for greater clarity. A histogram is a graphical representation, showing a visual impression of the distribution of the data. MS Excel has a useful histogram function. The important step is for the user to define the categories or 'bins' (discrete intervals). A histogram differs from a bar chart as it plots the intervals (bins) on the horizontal axis against the frequency of occurrence within each category along the vertical axis.

As an example we consider data collected by from active air-sample counts taken within a Grade B / ISO class 7 cleanroom. Over a period of time 824 samples are collected. By using descriptive statistics we have an idea that the data is of a low count (as expected within such an area): a mean count of 0.15 cfu, a modal count of 0 cfu and a media count of 0 cfu. The range of the data is from 0 to 7 cfu. For such a clean area we would expect the overwhelming majority of counts to be zero and for the distribution to be negatively skewed. To show this quickly and simply the data can be placed into a frequency table:

Bin*	Frequency	Cumulative %
0	754	91.50%
1	46	97.09%
2	10	98.30%
3	9	99.39%
4	2	99.64%
5	0	99.64%
6	2	99.88%
7	1	100.00%
8	0	100.00%
9	0	100.00%
10	0	100.00%
11	0	100.00%
12	0	100.00%
13	0	100.00%
14	0	100.00%
15	0	100.00%
More	0	100.00%

Table 8: Frequency data summary

* To construct a histogram from a continuous variable you first need to split the data into intervals, called bins.

And then represented graphically:

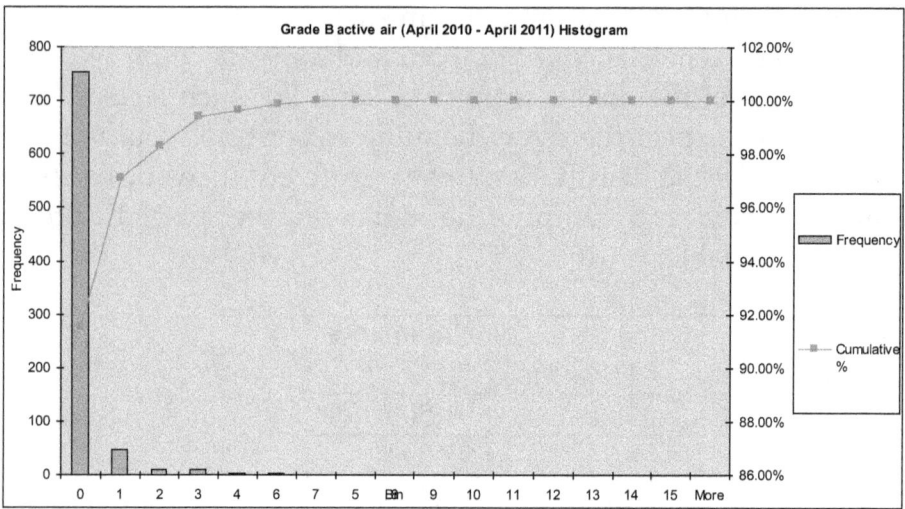

Figure 12: Histogram of frequency data

The histogram shows us that the data is distributed as we would expect from a Grade B / ISO class 7 cleanroom, with 91.5% of the results being zero and 100% of the results below the EU GMP action level of 10 cfu/m3. The advantage of histograms is that such data can be compared over time so that the proportion of counts being below the action level or as zeros can be trended to see if any shifts are occurring.

The histogram approach is also useful when reviewing the locations for environmental monitoring. Suppose, for example, that three active air-samples are taken within a cleanroom. If the microbiologist wished to remove one of the air-samples in order to save time and money, by constructing a histogram with all three samples and then constructing a histogram with one sample removed allows a comparison to be made. If the proportion of zero results does not significantly alter then there could be a case to be made for removing one of the air-samples.

For larger quantities of data and in order to show trends over time, control charting gives a clearer indication of the trend. Before examining the data, the theory behind control charts is discussed in greater detail.

Control Charts

A process may either be classified as in control or out of control. The boundaries for these classifications are set by calculating the mean, standard deviation, and range of a set of process data collected when the process is under stable operation. Then, subsequent data can be compared to this already calculated mean, standard deviation and range to determine whether the new data fall within acceptable bounds. For good and safe control, subsequent data collected should fall within three standard deviations of the mean.

The main purpose of using a control chart is to monitor, control, and improve process performance over time by studying variation and its source. There are several functions of a control chart (Nelson, 1985):

1. It centres attention on detecting and monitoring process variation over time.

2. It provides a tool for on-going control of a process.

3. It differentiates special from common causes of variation in order to be a guide for local or management action.

4. It helps improve a process to perform consistently and predictably to achieve higher quality, lower cost, and higher effective capacity.

5. It serves as a common language for discussing process performance.

Thus control charts are used to see if a process or set of data is in a state of statistical control (Alt, 1984: 110). If analysis of the control chart indicates that the process is currently under control (i.e. is stable, with variation only coming from sources common to the process) then data from the process can be used to predict the future performance of the process. If the chart indicates that the process being monitored is not in control, analysis of the chart can help determine the sources of variation, which can then be eliminated to bring the process back into control. For processes that are not in a state of statistical control, the charts will:

a) Show excessive variations;
b) Exhibit variations that change with time

The quality of the individual points of a subset is determined unstable if any of the following occurs (Steel and Torrie, 1980):

Rule 1: Any point falls beyond 3σ from the centreline (this is represented by the upper and lower control limits).

Rule 2: Two out of three consecutive points fall beyond 2σ on the same side of the centre line.

Rule 3: Four out of five consecutive points fall beyond 1σ on the same side of the centre line.

Rule 4: Nine or more consecutive points fall on the same side of the centre line.

When such conditions arise, control charts can be used to differentiate between such variations, to show:

i. Those variations that are normally expected of the process due chance or common causes (sometimes called 'random error'). These should be expected, to an extent;

ii. Those variations that change over time due to assignable or special causes. These often require some form of action.

The difference between common causes (a common cause variation is the 'noise' within the system) and special causes (new, unanticipated, emergent or previously neglected phenomena within the system, sometimes called 'systemic errors')) is important in determining what action to take for a follow up investigation. There is no consensus as to what a common cause is and what a special cause is. Some tentative examples include:

Common causes

- Inappropriate procedures
- Poor design
- Poor maintenance of machines
- Pipetting error
- Lack of clearly defined standing operating procedures
- Poor working conditions, e.g. lighting, noise, dirt, temperature, ventilation

- Substandard materials
- Contamination
- Bubbles in test tubes
- Ambient temperature and humidity
- With machinery (such as particle count generation from a filling machine): normal wear and tear or variability in machine settings
- Seasonal variations with data

Special causes
- Poor adjustment of equipment
- Poor batch of raw material
- Calibration errors
- Problem with a batch of test kits
- Power surges
- Change of process staff
- Change of culture media

This necessary investigative step will not be expanded upon in this book, but it does relate to the importance of microbiological data deviations and the assignment of corrective and preventative actions, which form a core part of pharmaceutical microbiology.

Control charts have three basic components:

1. A centre line (usually the mathematical average of the entire samples plotted).
2. Upper and lower statistical control limits. These are three standard deviations (or standard errors from the centre line).
3. The data plotted over a time period.

Thus, to start with, control charts require a target value. The target values used in the examples in this section related to data collected during 2003. Thus 2003 was used to establish the 'norm' for future samples in a given room. Using control chart terminology this was necessary to establish what the common variability of the environment was. This approach could be unrepresentative if 2003 did not represent a typical year. For the purposes of this section the reader will need to assume the use of 2003 as a benchmark was the correct approach to take.

Control charts use upper and lower control limits and warning limits, in addition to a straight line at the centre of the graph (which is the mean). Data is plotted against the central line. For most microbiological monitoring, normally only the upper levels are of concern (the upper control level: UCL, and the upper warning level: UWL respectively). These monitoring levels alert the chart user to react to major changes in trend. When limits are set using historical data it is expected that a certain percentage of results will fall outside the limits when plotted. The UCL and UWL can determine the probability (or 'chance') that future counts will fall outside.

In this context, warning limits represent approximately a 0.025 or 2.5% chance. Alternatively, sometimes a 5% chance is employed. The choice here is in reaching a balance between the probability of false alarm and the earlier detection of a possible problem. Control limits (or action limits) represent an approximate 0.001 or 0.1% chance and this is an approximation of three standard deviations from the mean. According to Klein (2000: 427), by employing upper limits set at three standard deviations, these are wide enough to avoid taking unnecessary action for what would otherwise prove to be 'statistical noise' (or the small rise and drop in low level microbial counts).

The detection of data rising above the UCL rise is theoretically rare (if the control chart has been set up correctly) and any results that would exceed it would indicate that:

i.   The variation is due to an assignable cause
ii. The process is out-of-statistical control (which may or may not be the same as out-of-microbiological control).

When a meaningful amount of data is plotted (such as one year or more of data), several points rising above the UCL are indicative of an out-of-control situation (and thereby this requires some form of corrective action). In contrast a single point outside of the control levels would be regarded as 'statistical noise'.

Types of control chart

There are a variety of different types of control charts. These different types are summarized in the table below:

Chart Type	Measures	Application
I-chart or X-chart	A control chart for processes in which individual measurements of the process are plotted for analysis. Also called an I-chart or X-chart.	Individual data points
X-Bar	A control chart used for processes in which the averages of subgroups of process data are plotted for analysis.	Means of data points
X-Bar and S	Process average and	High volume of

Chart Type	Measures	Application
	standard deviation	data, single characteristics. Sample size 2 or larger. For larger subsets, the range is a poor statistic to estimate the distributions of the subsets, and instead, standard deviation is used. In this case, the X-Bar chart will display control limits that are calculated using the average standard deviation.
X-Bar and R	Process average and range	High volume of data, single characteristics. Sample size between 2 and 6 (if larger than six, use an X-Bar and S chart). he X-Bar chart displays the centerline, which is calculated using the grand average, and the upper and lower control limits, which are calculated using

Chart Type	Measures	Application
		the average range. Future experimental subsets are plotted compared to these values. This demonstrates the centering of the subset values. The R-chart plots the average range and the limits of the range.
X and MR	Process average and moving range	Sensitivity not required Sampling is costly Long cycle time
CUSUM	Cumulative deviations from mean	Charts for individuals when X and MR are not sensitive enough.

A control chart designed to detect small process shifts by looking at the Cumulative SUMs of the deviations of successive samples from a target value. |
| EWMA | Weighted moving average | Charts for individuals when X and MR are not sensitive enough.

An Exponentially |

Chart Type	Measures	Application
		Weighted Moving Average control chart that uses current and historical data to detect small changes in the process. Typically, the most recent data is given the most weight, and progressively smaller weights are given to older data.

When setting up a control chart, it is important to establish the subgroup. Subgroup is another name for a sample from the population. From these subsets, a grand average, an average standard deviation, and an average range are calculated. The grand average is the average of all subset averages. The average standard deviation is simply the average of subset standard deviations. The average range is simply the average of subset ranges.

For example, in order to determine the upper (UCL) and lower (LCL) limits for the x-bar charts, you need to know how many subgroups (n) there are in your data. Once you know the value of n, you can obtain the correct constants (A2, A3, etc.) to complete your control chart. This can be confusing when you first attend to create a x-bar control chart. The value of n is the number of subgroups within each data point. For example, if you are taking temperature measurements every min and there are three temperature readings per minute, then the value of n would be 3. And if this same experiment was taking four temperature readings per minute, then the value of n would be 4.

If the subgroup size is not large enough, then meaningful process shifts may go undetected. On the other hand, if the subgroup size is too large, then chart signals may be produced from insignificant process shifts.

Examples in relation to microbiological data

The two analysis-charting tools used here as an example are:

1. A control chart called a cumulative sum chart (or cusum), which is the most suitable for large quantities of low count data collected over time and where only a small shift in data is anticipated.
2. A control chart called a Shewhart chart, which is more applicable for larger numbers and / or for detecting larger shifts in the data.

For both chart types, when considering shifts in data, a small shift in data can be regarded as between 0 – 0.5 standard deviations (sigma); whereas a larger shift in data is anything between 0.5 and 2.5 sigma.

Shewhart charts and cusum charts primarily differ by the way that they weight data. For Shewhart charts, the signal of the chart is based entirely upon the last plotted point (that is, each point bears no relation to the previously plotted point, or simply, the Shewhart chart has "no memory" and ignores immediate history). In contrast, the cusum chart is all based "on memory", where equal attention is given to the most ancient datum as the most recent. This can result in very different types of plots for the same data set and the chart user must be very clear what type of trend they wish to demonstrate before selecting one type of chart over another.

This section discusses both charts. The charts used in this section were prepared using a software package called Quality Analyst, manufactured by a company called Northwest Analytical, and using MS Excel. This book does not endorse any software package as being superior or inferior to others in the market, although the student may wish to consider software compatibility with other systems such as LIMS. As previously discussed, control charts typically require data to meet normal distribution, although this need not be too exact for many control charts are robust enough to deal with slight variations from normality (Chou, 2004: 241). Thus, before using a control chart, the data must be examined for 'normality'. One way to do this is to use a histogram or blob-chart and to qualitatively assess the data distribution. More sophisticated approaches involve the use of goodness-of-fit tests (such as Shapiro and Wilk's W test). For the following examples the student is to assume that this has taken place.

The cumulative sum chart

Cusum charts, introduced by Page in 1954, are used to decide whether a process is in statistical control by detecting a shift in the process mean. The cumulative sum chart is particularly useful for studying the gradual influence of time. Where there is no time trend, the cusum is flat. A change in the level over time is reflected in a change in the slope of the cusum (Everitt, 2003: 62; Chou, 2004: 244).

Cusums function by displaying cumulative sums of the deviations of measurements or subgroup means from a target value. This can be an increase or a decrease away from the target. Cusum charts are theoretically more sensitive to shifts in the process mean than Shewhart charts. Cusums will show:

i.   If changes have taken place;
ii.   Approximately when the change has taken place.

Cusum charts plot the cumulative sum of the deviations between each data point (a sample average) and a reference value, T (for target). Each point on a cusum chart is based on information from all samples (measurements) up to and including the current sample (measurement). Cusums measure the degree of variation from a target value (Hunter, 1986: 204).

When studying a cusum chart the main focus is with the slope of the plotted line, not just the distance between plotted points and the centreline. Significant variations from the target value are described as 'steps'. These can be in an upward or a downward direction. The concern for the type of data examined in this report is for upward steps because increasing microbial counts are of concern; for other applications, outside of reviewing microbial counts, the downward direction of charts maybe of interest also. It should be noted that the direction of the cusum can wander from the mean even when the data is generally on target.

Significant upward steps are when two out of the three values, from a subgroup that is plotted as a mean (one point on the graph), are outside of the target range. When this situation occurs the statistical software used here indicates this with a '*' sign to indicate a change in the trend (this can be an improvement or a potential out-of-control situation). These are descriptions of more significant events than simple upward steps.

The power of cusums can be increased through the use of V-masks. A V-mask is, unsurprisingly, in the shape of a V and is placed sidewise (>) with the vertex placed a fixed distance from the last plotted point. If the cusum graph is outside or moving outside of the V-mask, then the chart can be considered as going out of statistical control, whereas if the data falls within the V-mask branches it can be described as being within statistical control. If this movement is in a downward direction below the target line, this is indicative of the data having very low counts and being satisfactory. If the movement is in a upward direction and is above the target line, this is indicative of the data having a series of high counts and is unsatisfactory. The importance of V-masks is that one cusum can be compared to another, for the same data set, at different time periods. If the V-mask has shifted then an indication of the trend can be derived.

Worked Example: Grade B / ISO class 7 cleanroom

The following section uses an example of a Grade B / ISO class 7 clean room. This clean room was selected for studying using the cusum chart due to the expected low count data. Two types of sample are examined: contact plates and active air samples.

The target levels used were based on in-use action levels:

Action Levels (used to set control chart limits)

Room	Sample Type and action level			
Grade	CP (cfu)	SW (cfu)	AS (cfu)	SP (cfu)
B	2	2	3	2

Table 9: Action levels for the sample set.

umbers are cfu / unit of measurement.

Cusum target levels (adapted from Table 7)

These were set at 10% of the action level.

| Room | Sample Type and cusum target level | | | |
Grade	CP	SW	AS	SP
B	0.2	0.2	0.3	0.2

Table 10: Chart monitoring levels.

Key:

AS = air-samples
CP = contact plate
SW = swab
SP = settle plate

The two sample types from the data are examined. Brief explanatory text has been added below each chart to aid the interpretation.

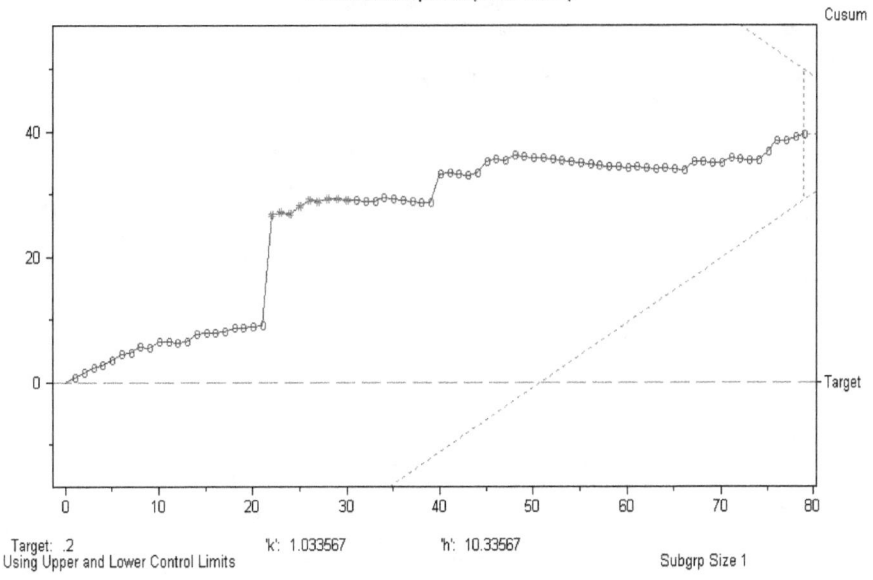

File: UNNAMED.DAT   26/4/105
File: UNNAMED.DAT
CP
P224 contact plates (2002 - 2004)

Figure 12: Contact plates from a Grade B / ISO class 7 cleanroom

Chart comments: two step upward trends took place one during the first quarter of the chart and one during the middle portion. The first upward step was of more significance, as indicated by the '*' symbol on the chart. This is indicative of seven or greater plotted points being of the same trend. However, the data then proceeded not to show any significant upward trend and remains in control, as indicated by the relation of the points to the V-mask.

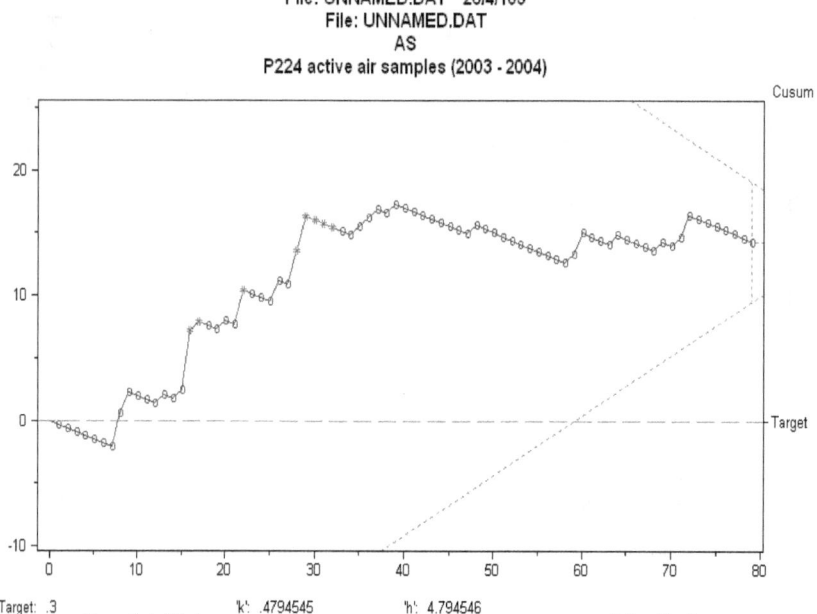

File: UNNAMED.DAT   26/4/105
File: UNNAMED.DAT
AS
P224 active air samples (2003 - 2004)

Figure 13: Active air-sample counts from a Grade B / ISO class 7 cleanroom.

Chart comments: The data shows an improvement on that collected during the first period. During the first period there were a series of significant steps directed away from the target value (as indicated by the '*' symbol on the chart). For the latter half of the chart, there were less active air samples that gave counts in succession and the trend is satisfactory as indicated by the placement of the V-mask. This accounts for the lower number of steps within the chart.

The mean counts, divided into categories for each calendar year, to which the data relates to are shown in the table below. These can aid the interpretation of the trend.

Sample type	2002 mean (cfu)	2003 mean (cfu)	2004 mean (cfu)
Contact plates	3.5	1.1	0.42
Active air	7.2	0.8	0.28

Table 13: Mean data counts examined over successive calendar years.

Interpretation: each sample type has shown a decrease in the mean count from 2002 to 2004. For each sample this has been a large reduction, with active air samples showing the greatest reduction by nearly 90%.

The Shewhart chart

The Shewhart chart is named after Walter A. Shewhart (1891-1967), a physicist at the Bell Telephone Laboratories, who introduced the method in 1924. The concepts underlying the control chart are that the natural variability in any process can be quantified with a set of control limits and that the variation exceeding these limits signals a change in the process (SAS Institute, 1999).

For biological and clinical applications, an adaptation of the Shewhart chart is used. This is called the Levey-Jennings chart. Both are sometimes called process-behaviour charts. An example of a Levey-Jennings chart is:

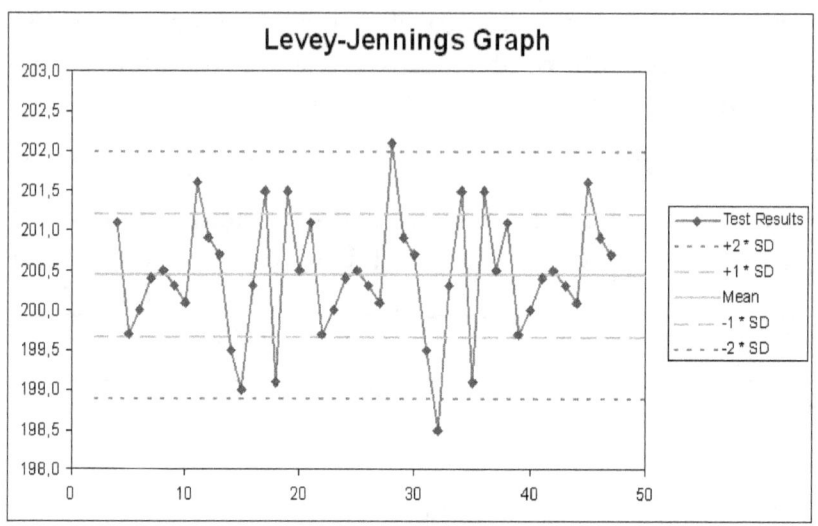

Figure 14: A Levy-Jennings chart

Shewhart charts either measure variables when the quality characteristic of a process are measured on a continuous scale (as in the examples used in this section) or they are used for attributes when the quality characteristic of a process is measured by counting the number of nonconformities (defects or high counts) in an item or the number of nonconforming (defective or high counts) items in a sample (this is the 'classic' SPC or 'statistical process control' chart). It should be noted that the underlying assumption is that the data is normally distributed. Therefore, it is important to understand (and to transform if necessary) the microbiological data before using control charts.

Charts typically consist of:

- Points representing a statistic (e.g., independent measurements, a mean, range, or proportion) of measurements of a quality characteristic in samples taken from the process at different times [the data].

These are plotted as a series of points on the control chart.

- The mean of this statistic using all the samples is calculated (e.g., the mean of the means, mean of the ranges, mean of the proportions)
- A centre line is drawn at the value of the mean of the statistic
- The standard error (e.g., standard deviation or square root (n) for the mean) of the statistic is also calculated using all the samples
- Upper and lower control limits (sometimes called "natural process limits") that indicate the threshold at which the process output is considered statistically 'unlikely' are drawn typically at 3 standard errors from the centre line.
- Some charts display limit values of ±1 standard deviation, ±2 standard deviations and ±3 standard deviations delineated on the chart.

The power of Shewhart charts is that they can be studied for unusual events or patterns in the data. Applications include:

- Gaining an understanding of the variability of the process
- Being able mark changes
- To determine if improved reactions occur following any changes
- To demonstrate effectiveness of any actions taken

Various events can be marked on the charts and the charts can be studied for changes in direction and for different 'movements'. These are sometimes called the Westgard Rules (named after James O. Westgard). Westgard rules are commonly used to analyse data in Shewhart control charts. Westgard rules are used to define specific performance limits for a particular assay and can be used to detect both random and systematic errors. These rules need to be applied carefully so that true errors are detected while false rejections are minimized.

Such movements and events include:

- A point lying beyond the control limits
- consecutive points lying beyond the warning limits
- 7 or more consecutive points lying on one side of the mean (this is an important issue because the 7th point is regarded as no longer being random)
- 5 or 6 consecutive points going in the same direction (indicates a trend)

Sometimes the rules are abbreviated (short hand notation) when charts are presented. For example the use of 1_{2s} to indicate 1 control measurement exceeding 2s control limits. Combinations of rules are generally indicated by using a "slash" mark (/) between control rules, e.g. $1_{3s}/2_{2s}$.

Shewhart charts are particularly useful in distinguishing variation due to SPECIAL CAUSES from variation due to COMMON CAUSES.

All charts require interpretation. This should be against a protocol in which the rules for the chart interpretation are defined (such as, which of the Westgard Rules will be applied). Some examples of general interpretation and response include:

- Determine the type of error based on your rule violation (random or systematic)
- Relate the type of error to the potential cause
- Inspect the testing process and consider common factors on multi-test systems

- Relate causes to recent changes
- Verify the solution and document the corrective action

In the context of environmental monitoring, special causes are local, sporadic problems such as the poor management of a particular water outlet in a process area or an air-sample count. In, general they are:

a) Localised.
b) Exceptions to the system.
c) Considered abnormalities.
d) Specific to a
e) Certain process;
f) Certain outlet or sampling site;
g) Certain method of sanitisation, etc.;
h) Sampling technique;
i) Equipment malfunction e.g. filling machine pumps;
j) Cross contamination in laboratory;
k) Engineering work;
l) Sanitisation frequencies.

Whereas common causes are problems inherent in the manufacturing system, such as a problem with the operation of a breakdown of the HVAC system where particle count control in a cleanroom has been lost. In, general they are:

- Inherent to the process because of:
- The nature of a system.
- The way a system is managed.
- The way a process is organised and operated.
- They can only be corrected by:
- Making modifications to a process.
- Changing a process for an alternative.

Once the special causes have been identified and eliminated, the process is said to be in STATISTICAL CONTROL *(OR STATISTICALLY STABLE)*. When statistical control has been established, Shewhart charts can be used to monitor the process for the occurrence of future special causes and to measure and reduce the effects of common causes (University of Newcastle-upon-Tyne, 2003).

Care needs to be taken in not confusing between common causes and special causes. To do so could result in a Type I or Type II error being made. A Type 1 Error is to treat as a special cause any outcome when it actually came from a common cause. Whereas, with a Type II Error something is to attributed a common cause to any outcome when it came from a special cause.

Once changes have been made the aim of monitoring using a Shewhart chart is to reduce variation in the plotted data.

Shewhart charts have control limits. Data is plotted against the control limits, where the centre line in the chart is the *grand average*. Shewhart charts also have an upper warning level and an upper control level, and corresponding lower levels, which are established around the process mean. The upper and lower control levels are approximations of three standard deviations from the process mean. There are different ways to calculate the control levels.

In the example below these were also based on the 2003 data, where it was expected that no more than 2.5% of the data would exceed the upper warning level and that no more than 0.1% of the data would exceed the upper control level. These represent an *approximation* for the microbial data because the assumption behind Shewhart charts is that the data is normally distributed and the data employed in the charts has been transformed (as previously discussed).

The target value for the Shewhart charts used in the example below were set based on 10% of the current action level. This was because the action level is statistically set so that approximately 10% of results are above the limit and 90% are below.

With the Levey-Jennings chart (mentioned earlier), a Levey-Jennings chart is a graph that quality control data is plotted on to give a visual indication whether a laboratory test is working well. It is named after S. Levey and E. R. Jennings who in 1950 suggested the use of Shewhart's individuals control chart in the clinical laboratory.

On the x-axis the date and time, or more usually the number of the control run, are plotted. A mark is made indicating how far off the actual result was from the mean (which is the expected value for the control).

The distance from the mean is measured in standard deviations (SD). Lines run across the graph at the mean, as well as one, two and sometimes three standard deviations either side of the mean. This makes it easy to see how far off the result was.

The Levey-Jennings chart differs from the Shewhart individuals control chart in the way that sigma, the standard deviation, is estimated. The Levey-Jennings chart uses the long-term (i.e., population) estimate of sigma whereas the Shewhart chart uses the short-term (i.e., within the rational subgroup) estimate.

Worked example: Grade C Clean Room

Similarly to the Grade B / ISO class 7 room examined above, two sample types have been examined. Brief explanatory text has been added below each chart to aid the interpretation.

Before plotting the data using a Shewhart chart the data was considered for a cusum chart. The data was not suitable for cusum charts due to the wide variation from the mean. For example, the contact plate data was shown to have a standard deviation of 1.19 and a variation of 1.42. To set target levels, data was plotted and the software was used to calculate the confidence intervals.

Action Levels

Room	Sample Type and action level			
Grade	CP (cfu)	SW (cfu)	AS (cfu)	SP (cfu)
C	21	2	15	33

Table 14: action levels applicable to the data

The control values for the Shewhart chart were set at 10% of the action level.

Room	Sample Type and cusum target level			
Grade	CP	SW	AS	SP
C	2.1	0.2	1.5	3.3

Table 15: monitoring levels applied to the Shewhart chart.

Contact plates

The confidence interval for the previous data was 8.11. This was used as the target value for the chart below:

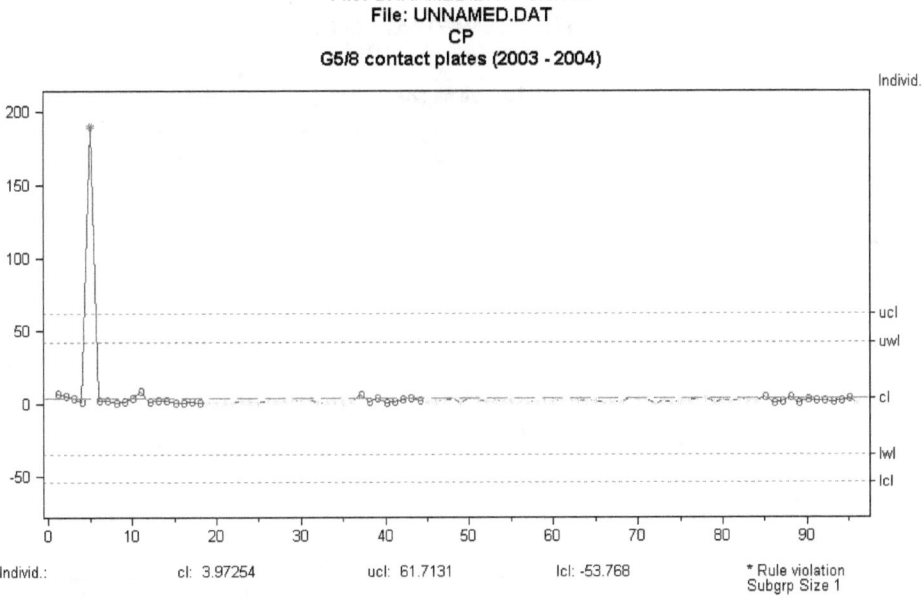

File: UNNAMED.DAT   9/5/105
File: UNNAMED.DAT
CP
G5/8 contact plates (2003 - 2004)

Figure 15: Contact plate data for a Grade C cleanroom

Chart comments: The start of the chart saw an increase in counts in the room, which accounted for a substantial upward step (as indicated by the '*' mark on the chart). However, this was shown to be an isolated result and one which did not cause a change in the trend. The results since showed a steady trend, and are without great variance from the target value. The trend is generally satisfactory.

Active air

The confidence interval for the previous data was 2.29. This was used as the target value for the chart below:

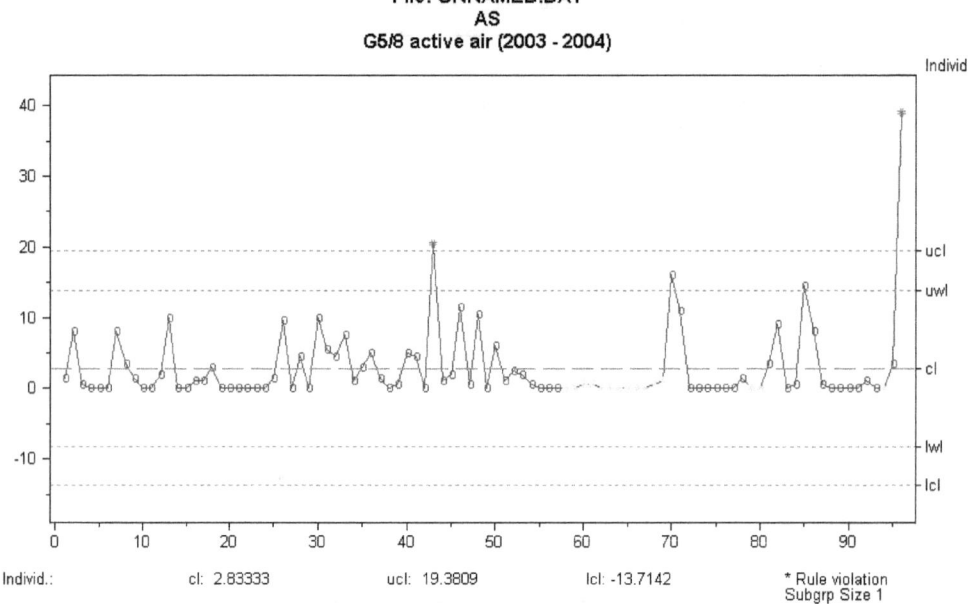

Figure 16: Active air-sample data from a Grade C cleanroom.

Chart comments: The chart indicates the counts during the latter part of the chart were satisfactory with most results clustered along the central line. Excursions above the upper warning limit were not sustained and no adverse trend developed. The final count for 2004 was very high and exceeded the upper control limit. However, there is no indication – based on the previous trend – that this would develop into a series of adverse counts.

The mean counts to which the data relates to are shown in the table below, and this can further aid the interpretation:

Sample type	2002 mean (cfu)	2003 mean (cfu)	2004 mean (cfu)
Contact plates	6.0	8.1	2.0
Active air	1.7	2.3	3.1

Table 16: Mean data counts examined over successive calendar years.

Interpretation: The contact plates have shown a decrease by approximately 4 cfu, and swabs have shown an increase, when data from 2004 is compared to that of 2003. The active air sample counts have increased by approximately +0.8 cfu. This was the second successive annual rise for these samples. However the trend was not adverse, as indicated by the control chart above.

Using MS Excel

The above examples used a statistical software package. As an alternative, Microsoft Excel can be used to create Shewhart charts. This uses the individual chart within Excel, with the addition of control levels. The calculation of the control level (CL), upper control level (UCL) and the lower control level (LCL) should be performed using this historical data. Statistical textbooks provide the required formula. The calculation is based on the assumption of normal distribution.

Table 17: manual calculation of control chart limits.

Setting Limits using control charts

For example, the Upper Control Limit:

99.8% UCL: $\mu \pm 3.09 \sqrt{\dfrac{\mu}{n}}$

And the Upper Warning Limit:

95% UWL: $\mu \pm 1.96 \sqrt{\dfrac{\mu}{n}}$

Once the chart parameters have been set up the chart can be plotted using a statistical package or a standard spreadsheet like MS Excel. Figure 10 illustrates such a plot by plotting data from a WFI system:

Figure 17: Excel developed Shewhart chart

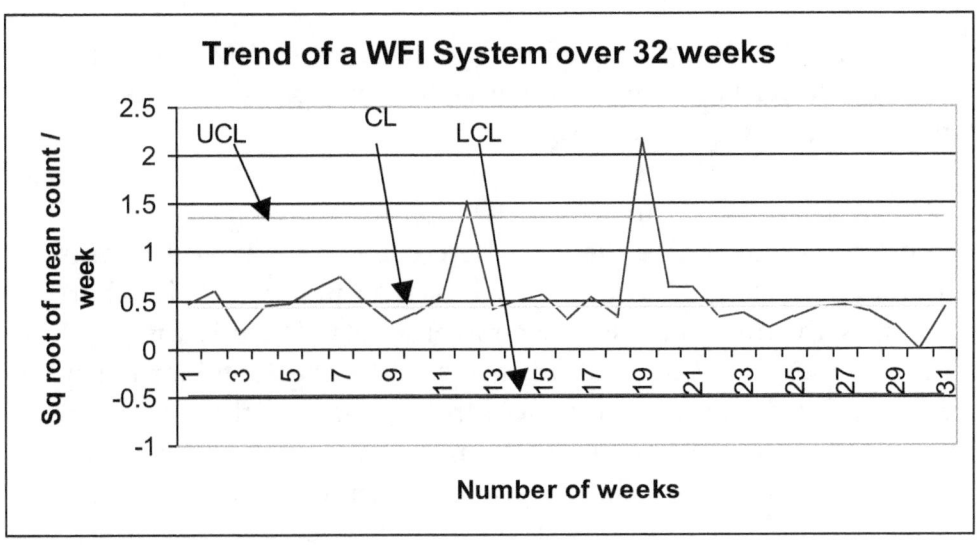

Where (**Table 18**):

UCL	Upper control level
CL	Control Level
LCL	Lower control level

Alternative approaches

An alternative approach to the use of the two control charts, for low count data, is to examine the intervals between counts rather than the intensity of the counts themselves. In such methods greater importance is placed on the direction of the trend (frequency intervals) rather than in rising or falling levels (chart magnitude). Such charts are useful to measure shifts in the average number of days between microbiological recoveries and are thereby useful for environments where few counts are expected (such as, at EU GMP Grade A / ISO class 5). For example, a chart could be set up to alert the user if the frequency of detecting a count moved from one every ten days to once every five days. Examples of such charts are Individual Value / Moving Range (I-MR) charts and

Exponentially Weighted Moving Average (EWMA charts), with the latter being more sensitive to small process shifts and the former to larger changes.

The EWMA chart can be used to plot a moving average of microbial recovery frequency where each of the past datum is assigned a weight. The charts looks similar to a Shewhart chart, with upper and lower control levels. The advantage of the EWMA chart is that it is insensitive to the normality assumption inherent in most control charts and thus data can be used without needing to transform it (Caputo and Huffman, 2004: 254-255). The chart also predicts what the next point is likely to be based on the previous data pattern. Alternatively it is possible to capture the principles of EWMA through frequency tables using descriptive statistics (see, for example, Cundell *et al*, 1998).

A further factor of the EWMA chart is that the weighting to previous data can be varied. This is in contrast to the Shewhart chart (no memory) and the cusum chart (maximum memory) in relation to previously plotted data. Thus the EWMA chart can be considered to fall half-way between a Shewhart chart and a cusum chart. If a Shewhart chart is considered as $\lambda=1$ and a cusum chart as $\lambda=0$, the EWMA chart can be set at any value between 0 and 1. According to Hunter (1986: 207), EWMA charts used in industry are normally are set at $\lambda=0.2 \pm0.1$. The smaller the value of λ, the greater the influence of historical data.

Discussion: control chart differences

This section has centred upon two types of control chart. Each chart type has advantages and disadvantages depending upon the type of data examined. These are summarised in the table below (**Table 19**):

Chart Type	Advantage	Disadvantage
Cumulative sum	Cusum charts are more sensitive to small process shifts.	abrupt shifts are not detected as fast as in a Shewhart chart.
Shewhart chart	Systematic shifts are easily detected.	The probability of detecting small shifts fast is rather small

This book could be criticised of over-simplifying the control chart approach. There are further complexities that arise when interpreting control charts, including the number of values that cross different control levels and whether the control levels should be re-assessed or re-calculated annually. These were not developed in this book. Furthermore, the charts used in this book are not the only approaches that can be taken. The most important point to make, however, is that while control charts are useful they are no substitute for professional judgement: a skilled microbiologist is still required to aid the interpretation of the control chart.

Computerised Systems

It is common for environmental monitoring data to be processed using computer software packages. Indeed, this is essential if sense is to be made of the mass of data assembled and has advantages over dealing with multiple manual paperwork entries, each of must be checked for transcription errors, which makes for a slow and tedious task!

Data can be gathered using a number of packages from simple Excel spreadsheets, through 'off-the-shelf' databases, to bespoke Laboratory Information Management Systems (LIMS). These systems allow data to be sorted, trended and for different graphs to be produced. Data can be examined over a variety of different time periods, including monthly, quarterly and annually. The time period over which data is examined is of importance. Producing data too often obscures the true picture and too often means that problematic trends can be missed.

When setting up data systems for use in the pharmaceutical industry a number of regulations must be considered. This is in addition to considering the many problems associated with microbiological environmental and utility monitoring data control, documentation, reporting and trending. The failure to adequately deal with these areas are among those most commonly cited as objectionable during FDA inspections of manufacturing facilities (from a review of inspectorate trends in 2005).

The FDA introduced critical regulations relating to electronic storage on 20th August 1997. The FDA regulation covers the all information relating to the following:

"Electronic records, electronic signatures, and handwritten signatures executed to electronic records as equivalent to paper records and handwritten signatures executed on paper" (Reference: Title 21 Code of Federal Regulations (21 CFR Part 11)).

Together with European inspectorate concerns, the most important regulations include:

- Title 21 Code of Federal Regulations FDA Part 11 compliance (especially relating to Electronic Record Keeping)
- IPSE Good Automated Manufacturing Practices (GAMP)
- EU GMP Annex 11, 'Computerized Systems'
- EN 60601-1-4 Medical Electrical Equipment

These regulations have a number of things in common. This includes guidance relating to the:

- Retention of electronic records.
- The need for secure and reliable password-protected databases.
- The importance in tracking, logging and time stamping data.
- The need for extensive audit trails for records, particularly in indicating the date and time when electronic records are created, modified, or deleted; also, the operator identification,; recording prior values for changed items and the reason for making a change. It is important to demonstrate to regulatory agencies that any possibility of falsification of any records has been avoided.
- The regulations also indicate the importance of validation. New electronic systems require the now

'standard' validation steps of Installation Qualification; Operational Qualification; and Performance Qualification. Any system may be open to regulatory audit and review. All validation must follow the 'life cycle' approach (including verification and validation).

The consideration of software issues has been relatively brief, the reader is advised to explore this subject further as necessary.

Case Study: Trending data from a microbiological water system

Microbiological testing of pharmaceutical water systems, like environmental monitoring, generates a mass of data that can be difficult to sort, analyse and interpret. The microbiologist will set alert and action limits based on historical data, which will enable out of limits (OOL) results to be detected, investigated and for corrective and preventative action to be formed.

However, the simple detection of results exceeding action levels is often a case of 'shutting the door after the horse has bolted'. For microbiological analysis, the detection of the trend *before* results exceeding action levels is far more important. The examination of the direction of a trend (particularly those which are upward or adverse) allows the organisation to take action before various outlets require quarantining or the water generation plant requires shutting down. With water systems, the value of trend analysis is further reinforced by the relatively long time taken to obtain a result. The widespread use of the culture medium R2A has increased the time taken to obtain results to anything up to fourteen days. Therefore, the microbiologist needs to be able to take action well in advance of the OOL result being read.

The other advantage of performing trend analysis is to gain an understanding of the particular water system: how it behaves during times of high and low usage; seasonal variations; the effect of time between routine system sanitisastions and so on.

Results from the microbiological monitoring of water systems generate a large amount of information. Various different approaches can be used to make sense of this data.

The first question in designing a trend analysis scheme is: what to trend? Water systems require examination over time in order to detect movement or to retrospectively study any seasonal variations. A typical system will have between twenty and sixty user outlets. To plot each outlet individually would, if there was sufficient time to do this, would not give a full understanding of the system. The system's pattern can be revealed from a study of the mean count obtained from the system for a given day or a week is the most straightforward approach. This is the basis of the examples used below. The approach taken may be required to individual outlets or different sub-loops of the system.

The approach to trending examined in this section is similar to the approaches used for the environmental monitoring data above. The reader is advised to refer to environmental monitoring section where necessary.

Control charts

As discussed earlier, control charts are particularly useful in distinguishing variation due to SPECIAL CAUSES from variation due to COMMON CAUSES.

In relation to water systems and testing, special causes are local, sporadic problems such as the poor management of a particular water outlet in a process area. In, general they are:

a) Localised

b) Exceptions to the system

c) Considered abnormalities

d) Specific to a

a) A certain process

b) A certain outlet

c) A certain method of sanitisation, etc.

Whereas causes are problems inherent in the manufacturing system, such as a problem with the operation of a water storage tank feeding the outlets on the distribution loop. In, general they are:

a) Inherent to the process because of:

  i)   The nature of the system
  ii) The way the system is managed
  iii) The way the process is organised and operated

b) They can only be removed by

  i)   Making modifications to the process
  ii)   Changing the process

Once the special causes have been identified and eliminated, the process is said to be in STATISTICAL CONTROL *(OR STATISTICALLY STABLE)*. When statistical control has been established, control charts can be used to monitor the process for the occurrence of future special causes and to measure and reduce the effects of common causes.

Application of Control Charts to Microbiological Analysis

Shewhart charts have two basic assumptions: first, that each plotted point is independent and secondly, that the data analysed is normally (binomially) distributed. How do these factors relate to microbiological data? For the former, the values plotted are from different samples, weekly means and so on; these are thereby independent of each other. Therefore, the first assumption is applicable to microbial counts from water systems.

For the second point, however, microbiological data is unlikely to be normally distributed (as represented by the 'classic' bell-shaped curve). This is a problem because control charts approximate a series of normal distribution plots over time. Normal distribution is a phenomenon found in many aspects of physical and biological science (from measurements like length to human height). However, micro-organisms are rarely (if ever) normally distributed. In a WFI system, for instance, the microbial counts will be predominately zero with a decreasing number of counts.

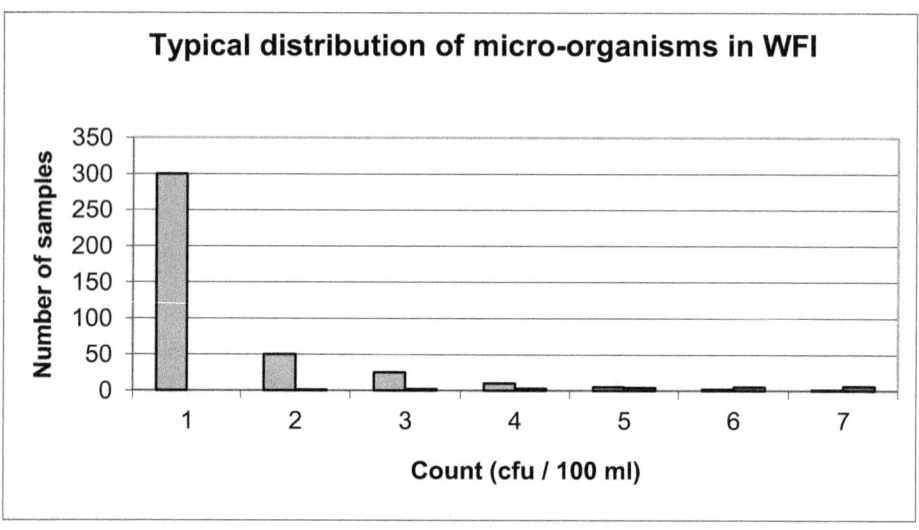

Figure 18:The chart above shows the typical distribution of micro-organisms in WFI.

Furthermore, micro-organisms distributed in water systems invariably follow Poisson distribution in that one sample from a water system may contain a large number of micro-organisms, whereas a second consecutive sample may contain none. Imprecision in sampling also add to this effect. For

Poisson distribution the frequency of counting 'events' over 'time' is more random.

The phenomena of Poisson distribution accounts for events where a sample from a water system may exceed an action level on one day, be below it for another two days and then be above it again. This situation does not indicate contamination appearing and disappearing, or that one sample has given the correct result and the other an unrepresentative one.

How do these observations apply to Control Charts?

Firstly, data from a water system should first be examined to see if it follows normal distribution before embarking on further analysis. Although, as stated earlier, it is improbable that the distribution of micro-organisms in water will follow normal distribution, such analysis remains the more accurate approach so it is incumbent upon the analyst to demonstrate if there is such distribution.

An example of such analysis is illustrated below. Table 20 summarises the mean count from a Water for Injection system for a period of twenty weeks.

Table 20: Data from a WFI System

Week No.	Mean count per week (cfu / 100 ml)
1	0.00
2	5.15
3	0.29
4	6.93
5	1.86
6	1.47
7	0.10
8	0.00
9	2.22
10	3.95
11	0.11
12	1.25
13	0.00
14	6.34
15	0.31
16	0.45
17	2.70
18	0.89
19	0.65
20	3.45

If the data is plotted, the distribution is:

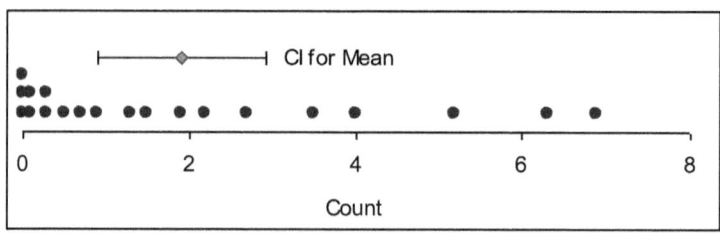

Figure 19: Blob plot of table data

The table and distribution plot indicate that the results show a positively skewed distribution (as opposed to normal distribution), with the majority of the results of low count and a few higher counts to the extreme.

The data as it stands is not suitable for analysis by a control chart otherwise gross distortions would occur. However, the data can be *transformed* into an approximation of normal distribution and thereby plotted onto a control chart. There are no pre-set 'rules' for data transformation. A common approach, as outlined by Sokal and Rohlf, is to transform low count data (such as that obtained from a WFI system) by taking the square root; and to transform data with higher counts by converting it to logarithms at base 10 ($\log_{10}$). In performing $\log_{10}$ transformations, in the event where the mean count is zero, this requires the addition of a value of '1' to each value of zero. So that this does no distort other results, a value of '1' is added to each item of data.

Trend analysis of WFI

To transform low count data, such as WFI, the square root of the mean count for the week is taken. This is illustrated in Table 21 where the data examined earlier has been converted.

Table 21: Data from a WFI System (transformed)

Week Number	Mean count per week (cfu / 100 ml)	Square root of mean
1	0	0.00
2	5.15	2.27
3	0.29	0.54
4	6.93	2.63
5	1.86	1.36
6	1.47	1.21
7	0.1	0.32
8	0	0.00
9	2.22	1.49
10	3.95	1.99
11	0.11	0.33
12	1.25	1.12
13	0	0.00
14	6.34	2.52
15	0.31	0.56
16	0.45	0.67
17	2.7	1.64
18	0.89	0.94
19	0.65	0.81
20	3.45	1.86

A distribution diagram of the transformed data from this table shows something closer to normal distribution:

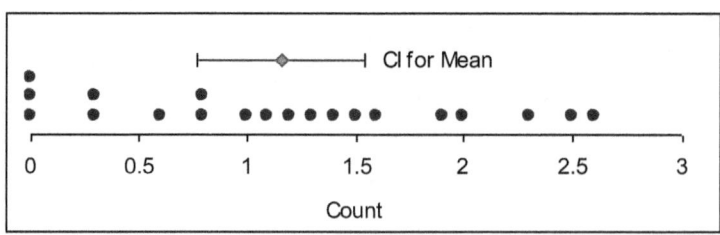

Figure 20: Blob plot of table data

The first step in constructing a control chart requires the analysis of a preliminary set of data that is assumed to be in a state of statistical control. This analysis, Phase I analysis, is conducted to estimate parameters that will be used subsequently for on-going monitoring of the process or Phase II . The distinction between these two phases of data collection is important. The first phase of data collection should utilize a very large sample of data, such as one year, so that parameters and control limits are estimated for Phase II. By taking one year of data, special cause fluctuations will be minimised. The analyst may wish to consider removing extreme counts ('outliers') from a particular week *if* an appropriate statistical tool is applied and if the removal of the counts can be justified as a *special cause* variation and is clearly not a common cause variation (such as Barnett and Lewis, 7).

The calculation CL, UCL and LCL should be performed using this historical data. If a new water system is being monitored, either tentative limits should be established or the new system could be plotted using the limits from the system being replaced. Statistical textbooks provide the required formula or this can be expedited with statistical software (or see further 5). The calculation is based on the assumption of normal distribution.

Table 22: Data for control charts

These control limits need to be re-calculated on a frequent basis (such as annually). This is because systems can change profile over time.

Once the chart parameters have been set up the chart can be plotted using a statistical package or a standard spreadsheet like MS Excel. Figure 26 illustrates such a plot:

Figure 26:

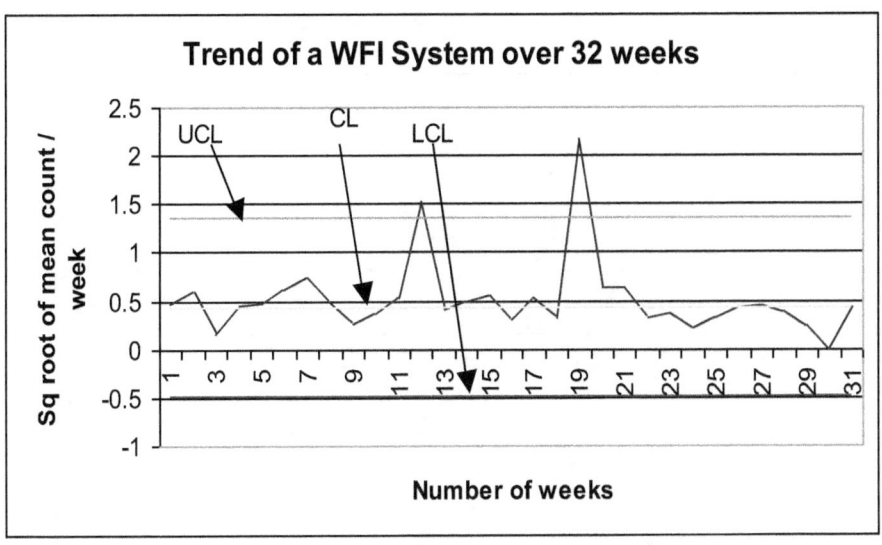

Where:

UCL	Upper control level
CL	Control Level
LCL	Lower control level

The above graph shows the mean count from a WFI System, for a series of weeks. The mean count has been transformed through taking the square root of the mean count for each week. This is in order to approximate normal distribution. The data has been plotted, and rises and falls, along the control level (the central flat line). The movement of the data is examined against the upper control level (the highest placed of the flat lines). Where results have exceeded the control level, an explanation is required.

Once the graph has been set up, it can be added to at frequent intervals so that the trend can be discerned. Figure 27 displays the same chart with an additional six months data:

Figure 27:

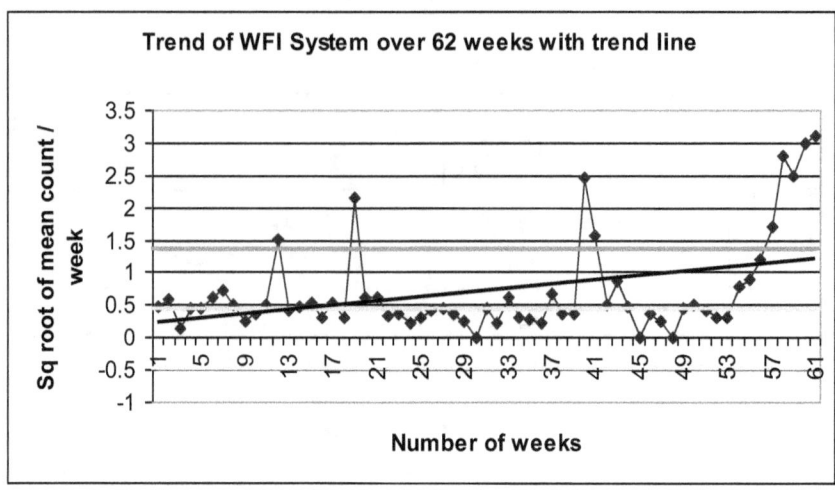

The charts indicate the start of an adverse trend, via the addition of a linear plot trend line, and out-of-control situation, with a series of points rising above the upper control level. In the example chart (in Figure 27) the organisation should have taken action concerning the water system and have closed the system down for investigation and formulation of appropriate preventative action.

<u>Trend analysis of purified water</u>

The analysis of the second water system is very similar. The key difference is the method of data transformation. This is due to the high level of microbial counts.

To transform higher count data, such as results from a purified or purified water system, a value of '+1' is added to each mean count, and the $\log_{10}$ of the mean count + 1 for the week is calculated. This is illustrated in Table 28 using some example data.

Table 28: Data from a purified water system

Week Number	Mean Count of week	Mean Count +1	Log_{10} of mean count
1	1110	1111	3.05
2	2347	2348	3.37
3	789	790	2.90
4	2350	2351	3.37
5	1098	1099	3.04
6	6590	6591	3.82
7	2104	2105	3.32
8	2673	2674	3.43
9	860	861	2.93
10	908	909	2.96
11	1103	1104	3.04
12	1079	1080	3.03
13	580	581	2.76
14	999	1000	3.00
15	876	877	2.94
16	1045	1046	3.02
17	3010	3011	3.48
18	2987	2988	3.48
19	2589	2590	3.41
20	2077	2078	3.32
21	2167	2168	3.34
22	1009	1010	3.00
23	1469	1470	3.17
24	1550	1551	3.19
25	1444	1445	3.16
26	1467	1468	3.17
27	1440	1441	3.16
28	1001	1002	3.00
29	797	798	2.90
30	682	683	2.83

The calculation of the confidence intervals is as per the WFI example, and the need to re-assess these chart parameters remains.

Two charts (Figures 28 and 29) show the changing profile of a purified water system over time.

Figure 28:

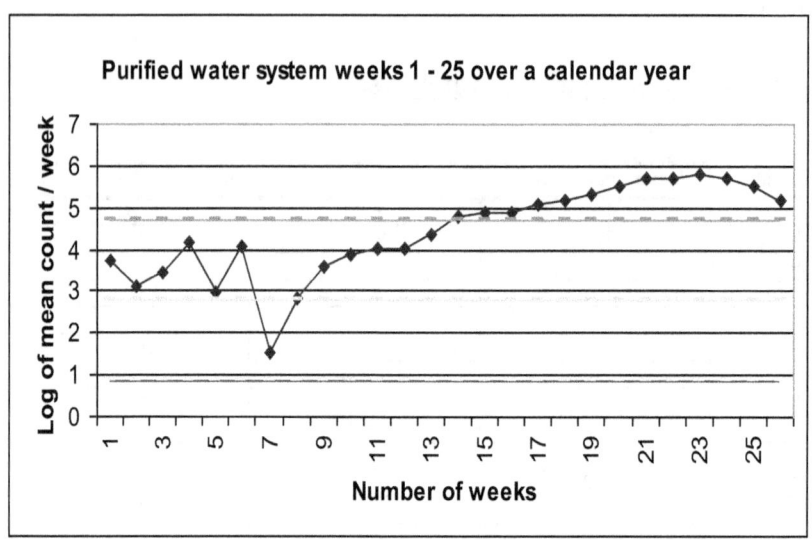

Purified water system weeks 1 - 25 over a calendar year

Figure 29:

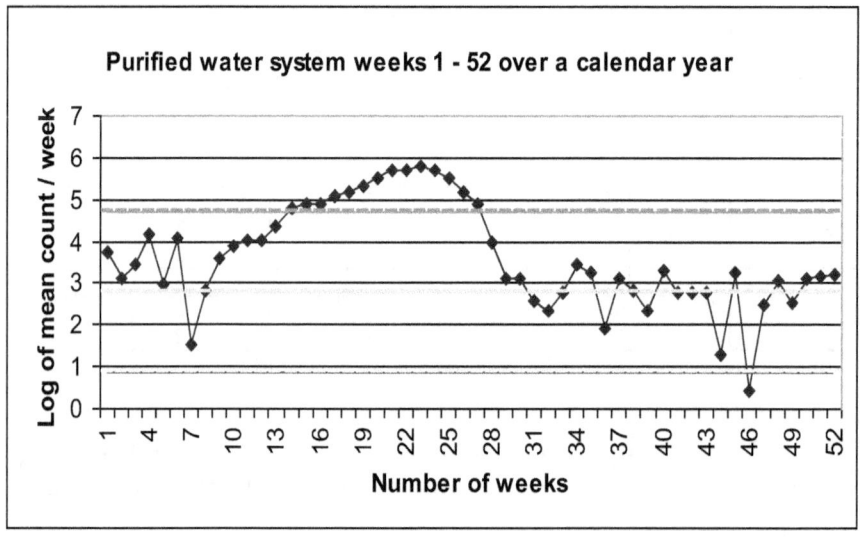

Purified water system weeks 1 - 52 over a calendar year

Figure 28 shows the development of an out-of-control situation. Necessary corrective action would have been taken at the time, from an alert given by the microbiologist to a multi-disciplinary team.

Figure 29 allows greater consideration of preventative action whereby seasonal variations can be examined. The high peak in this chart occurred during the summer months (from May to September) and can indicate that the water plant is at greater risk at this time of year.

If two year's worth of data are considered, as in Figure 30, this seasonal affect is most apparent and sends a clear signal to the organisation where they need to focus their activities (for example, increasing water sanitisaton frequencies).

Figure 30:

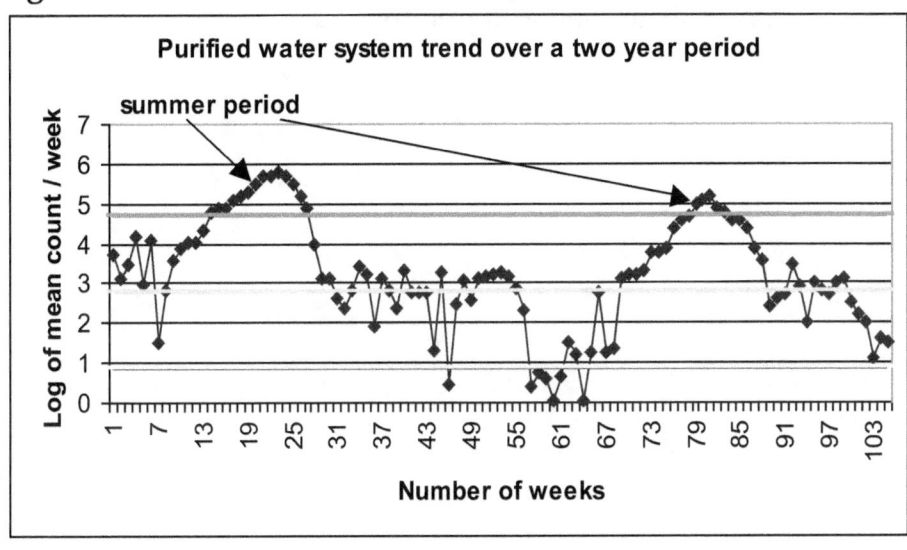

Summary

This section has outlined a possible approach for the examination of microbiological data from two different water systems (WFI and purified water).

There are different approaches that can be taken. This paper has explored just one. No single approach is statistically valid or invalid. However, where common statistical tools are used the fact that they are inappropriately designed for data that is normally distributed must be taken into account. It is often necessary to mathematically transform such data before commencing statistical analysis. The most important aspect in selecting an approach is to remain consistent so that data can be meaningfully assessed over time and that your water system can be examined and understood.

References and Further Reading

Ackers, J. (1997): 'Environmental monitoring and control', <u>PDA Journal of Pharmaceutical Science and Technology</u>, Vol. 51, No.1

Ackers, J. and Agallaco, J. (2001): 'Environmental Monitoring: Myths and Misapplications', <u>PDA Journal of Pharmaceutical Science and Technology</u>, Vol. 55, No.3, pp176-184

Alt, F.B. (1984): 'Multivariate Quality Control' in Koltz, S. and Read, C. B. *The* <u>Encyclopaedia of Statistical Sciences</u>, 1st Edition, John Wiley and Sons, pp. 110-122.

Barnett, V. and Lewis, V. (1995): '<u>Outliers in Statistical Data</u>', John Wiley and Sons, 3rd Edition

Black, J.G. (1996). *Microbiology. Principles and Applications.* Third Edition. Prentice Hall. Upper Saddle River, New Jersey. pp. 136-140

Casella G. Berger RL (2001). *Statistical Inference.* Duxbury Press

Caputo, R.A. and Huffman, A. (2004): 'Environmental Monitoring: Data Trending Using a Frequency Model', <u>PDA Journal of Pharmaceutical Science and Technology</u>, Vol. 58, No.5, pp254-260

Cochran, W.G. (1977): *Sampling Techniques,* 3rd edition, J Wiley & Sons: New York

Chou, C.J. (2004): 'Groundwater Monitoring: Statistical Methods for Testing Special Background Conditions' in Wiersma, G.C. (ed.): '<u>Environmental Monitoring</u>', CRC Press, Boca Raton, pp239-256

Cundell, A. M. (2004): 'Microbial Testing in Support of Aseptic Processing', *Pharmaceutical Technology*, June 2004, pp56-64

Cramer, D. and Howitt, D. (2004): *The Sage Dictionary of Statistics*, Sage, London

Cundell, A.M., Bean, R., Massimore, L. and Maier, C. (1998): 'Statistical Analysis of Environmental Monitoring Data: Does a Worst Case Time for Monitoring CleanRooms Exist?', <u>PDA Journal of Pharmaceutical Science and Technology</u>, Vol. 52, No.6, pp326-330

Deschenes, P. 'Viable environmental monitoring', in Carleton. F. And Agalloco, J. (eds.) '<u>Validation of pharmaceutical processes</u>', Marcel Dekker Inc. New York

Eisenhart, C. and Wilson, P.W. (1943): 'Statistical Methods and Control in Bacteriology', *Microbiology and Molecular Biology Reviews*, Vol. 7, No.2

Everitt, B. S. (2003): *Medical Statistics from A to Z*, Cambridge University Press, Cambridge

Everitt, B. (2006): *Medical Statistics from A to Z*, 2nd edition, Cambridge University Press, Cambridge

Everitt, B.S. and Palmer, C.R. (2011): *Encyclopaedic Companion to Medical Statistics*, Wiley: Chichester, UK

Hetroys, P. *et al* (1997): 'Moving towards an microbiological environmental monitoring programme', <u>PDA Journal of Pharmaceutical Science and Technology</u>, Vol. 51, No.1

Hewitt, W. (2004): *Microbiological Assay for Pharmaceutical Analysis: A Rational Approach*, Interpharm: Boca Raton

Heyes, S. *et al* (1988): *Starting Statistics*, Weidenfield and Nicholson, London

Fleming, M.C. and Nellis, J. G. (1994): 'Principles of Applied Statistics', Routledge, London
Hinton, P.R. (1996): *Statistics Explained*, Routledge, London

Ilstrup. D.M. (1990): 'Statistical Methods in Microbiology', *Clinical Microbiology Reviews*, Vol.3, No.3, pp219-226

Jimenez, L. (2004): 'Rapid Methods for Pharmaceutical Analysis' in Jimenez, L. (Ed.) *Microbial Contamination Control in the Pharmaceutical Industry*, Marcel Dekker, New York, pp147-182

Klein, M. (2000): 'Two Alternatives to the Shewhart $\bar{X}$ Chart', Journal of Quality Technology, 32, 4, pp. 427-431

Kourti, T. and MacGregor, J.F. (1996): 'Multivariate SPC Methods for Process and Product Monitoring', Journal of Quality Technology, 28(4), pp. 409-428.

Lovegrove-Saville, P. and Perry, M. (2000): 'Setting environmental alert and action limits', *Pharmig News*, Number 3

Madigan, M. and Martinko, J. (editors) (2006). *Brock Biology of Microorganisms* (11th edition). Prentice Hall

Nelson, L. S. (1985), Interpreting Shewhart X Control Charts, *Journal of Quality Technology*, 17:114-16.

Paulson, D. S. (2008): *Biostatistics and Microbiology: A Survival Manual*, Springer, London.

PDA Technical report Number 13 (Revised), 2001 (Supplement to Vol. 55, No.5 of the PDA Journal of Pharmaceutical Science and Technology)

Prabhu, S.S. and G.C. Runger (1997): Designing a Multivariate EWMA Control Chart, Journal of Quality Technology, **29**(1), pp. 8-15.

Richter, Steven. 1999. Product Contamination Control, a Practical Approach Bioburden Testing. *J Validation Technology*. 5:333-336

Statistics for Microbiology (2001), *Statistics for Industry*, Knarlesborough

Stearman, R. (1955): 'Statistical Concepts in Microbiology', *Microbiology and Molecular Biology Reviews*, Vol. 19, No.3

Steel, R. G. D. and J. H. Torrie (1980), *Principles and Procedures of Statistics*, New York: McGraw-Hill.

Sokal, R.R. and Rohlf, F.J. (1995): 'Biometry: The Principles and Practice of Statistics in Biological Research', W. H. Freeman and Co., New York, pp411-422

Tang, S. (1998): 'Microbial Limits Reviewed: The basis for unique Australian regulatory requirements for Microbial Quality of Non-Sterile Pharmaceuticals', PDA Journal of Pharmaceutical Science and Technology, Vol. 52, No.3, pp100-109

Tetzlaff, R. 1992. Investigational Trends: Clean Room Environmental Monitoring. PDA J Parenteral Sci Tech. 46:206-214

Tortora, G.J., Funke, B.R., Case, C.L. (1995). Microbiology. An Introduction. Fifth Edition. The Benjamin/Cummings Publishing, Co., Inc., Redwood City, CA, pp. 155-158.

University of Newcastle-Upon-Tyne (2003): 'SPC Tools – Control Charts, Chemical Engineering Department, University of Newcastle-Upon-Tyne at http://lorien.ncl.ac.uk/ming/spc/spc8.htm

Wilson, J. (1997): 'Setting alert / action levels for environmental monitoring progams', PDA Journal of Pharmaceutical Science and Technology, Vol. 51, No.4

Wilson, J. (2001): 'Environmental Monitoring: Misconceptions and Misapplications', PDA Journal of Pharmaceutical Science and Technology, Vol. 55, No.3, pp185-190

www.ingramcontent.com/pod-product-compliance
Lightning Source LLC
Chambersburg PA
CBHW051507170526
45166CB00001B/424